Microprocessor and Microcomp

Macmillan Basis Books in Electronics
Series Editor Noel M. Morris

Beginning Basic P.E. Gosling
Continuing Basic P.E. Gosling
Digital Electronic Circuits and Systems Noel M. Morris
Electronic Circuits and Systems Noel M. Morris
Linear Electronic Circuits and Systems G.D. Bishop
Microprocessors and Microcomputers Eric Huggins
Semiconductor Devices Noel M. Morris
Microprocessor and Microcomputer Technology Noel M. Morris

Other Macmillan Books of Related Interest

Digital Techniques Noel M. Morris
Program Your Microcomputer in Basic P.E. Gosling

Microprocessor and Microcomputer Technology

Noel M. Morris

Principal Lecturer
North Staffordshire Polytechnic

First published 1981 by
THE MACMILLAN PRESS LTD
London and Basingstoke
Companies and representatives throughout the world

ISBN 0 333 32005 0 (hard cover)
ISBN 0 333 29268 5 (paper cover)

Printed in Hong Kong

Typeset in 10/12 Press Roman by
MULTIPLEX techniques ltd, Orpington, Kent.

Contents

Foreword

1 An Introduction to Microprocessors **1**

1.1	Digital computers	1
1.2	Hardware, software and firmware	4
1.3	A basic microcomputer	4
1.4	Bits, nibbles and bytes	6
1.5	Input/output ports	7
1.6	A simple microprocessor with I/O ports	9
1.7	A simple I/O sequence	11
1.8	Conclusion	14
Problems		

2 Binary Numbers and Arithmetic **15**

2.1	Numbering systems	15
2.2	Converting a number of any radix into its decimal equivalent	18
2.3	Converting a decimal number into its equivalent in any radix	19
2.4	Converting a binary number into an octal number	20
2.5	Converting a binary number into a hexadecimal number	20
2.6	Binary-decimal codes	20
2.7	Multi-number representation in BCD	22
2.8	Error detection in binary code groups	22
2.9	Alphanumeric codes	23
2.10	Binary addition	24
2.11	Hexadecimal addition	26
2.12	Binary complement notation	26
2.13	Binary subtraction	29
2.14	Hexadecimal subtraction	30
2.15	Multiword binary arithmetic	31
2.16	BCD (decimal) addition	32
2.17	BCD addition in microprocessors	33
2.18	Binary multiplication	34
2.19	Binary division	36
Problems		38

3 Logic Devices **40**

3.1 Truth tables 40
3.2 Logic signal levels 41
3.3 Logic gates 42
3.4 The AND gate 42
3.5 The OR gate 43
3.6 The NOT gate 45
3.7 The NAND gate 45
3.8 The NOR gate 46
3.9 Universal gates - NAND and NOR gates 48
3.10 Circuit implementation using NAND gates 49
3.11 Circuit implementation using NOR gates 50
3.12 The EXCLUSIVE-OR gate 50
3.13 Flip-flops, latches and flags 51
3.14 Integrated circuits 56
3.15 Three-state logic gates 57
3.16 Decoders 60
3.17 A practical I/O port 62
3.18 Basic I/O connections to a microprocessor bus system 64
3.19 Priority encoders 66
Problems 67

4 A Simple Microcomputer System **70**

4.1 The bus system 70
4.2 Architecture of a basic microprocessor 75
4.3 A simple microprocessor with I/O facilities 77
4.4 A simple program 80
4.5 The unconditional jump (JMP) instruction 84
4.6 Control bus signals 85
Problems 86

5 Requirements of a Microcomputer System **87**

5.1 A practical system 87
5.2 Monitor program 91
5.3 Keyboard scanning 91
5.4 Light-emitting diode (LED) displays 97
Problems 102

6 Memory Organisation **103**

6.1 The need for ROM and RAM 103
6.2 A basic read-only memory 104
6.3 Mask programmed read-only memory (ROM) 106

6.4 Field-programmed ROM (PROM) 106
6.5 Erasable PROM (EPROM) 107
6.6 Comparison of ROM, PROM and EPROM 108
6.7 Electrical connections to the ROM 110
6.8 Random-access memories (RAM) 112
6.9 Static RAMs 112
6.10 Dynamic RAMs 116
6.11 Memory expansion 117
6.12 Memory maps 120
6.13 Relationship between the size of RAM and ROM 127
6.14 Direct memory access (DMA) 127
Problems 127

7 Input/Output **129**

7.1 Input/output ports 129
7.2 Isolated I/O or accumulator I/O device selection 130
7.3 Memory-mapped I/O 134
7.4 Attached I/O or on-chip I/O 136
7.5 Handshaking 136
7.6 A programmable I/O port containing a memory 138
7.7 Organisation of a typical programmable RAM I/O port 139
7.8 An application of a programmable I/O port 140
7.9 Serial data transmission 147
7.10 Using shift registers as serial I/O devices 151
7.11 Interface standards 151
Problems 155

8 Subroutines and Stacks **156**

8.1 The need for subroutines 156
8.2 The need for a stack 157
8.3 Stack organisation 158
8.4 Use of the stack for data and status storage 161
8.5 A time delay subroutine 164
8.6 Nested subroutines 168
8.7 Subroutine calls using the RESTART instruction 169
8.8 Conditional CALL and RETurn instructions 171
Problems 172

9 Interrupts and Polling **173**

9.1 The meaning of polling and interrupts 173
9.2 Basic features of an interrupt system 175
9.3 Types of interrupt 175

9.4 Vectored interrupts 176
9.5 Interrupt methods 177
9.6 Vectored interrupt hardware 178
9.7 Interrupt program organisation 182
9.8 Priority interrupts 183
9.9 Timers 185
Problems 186

10 A Typical Instruction Set **187**

10.1 Types of instruction 187
10.2 Instruction and data format 187
10.3 Microinstructions and microprograms 188
10.4 Macro-assemblers and pseudo-operations 191
10.5 Addressing modes 191
10.6 Direct addressing 192
10.7 Indirect addressing 192
10.8 Immediate addressing 193
10.9 Indexed addressing 193
10.10 Relative addressing 194
10.11 Register direct addressing 197
10.12 Register indirect addressing 197
10.13 Stack addressing 198
10.14 Instruction categories 198
10.15 Symbols and abbreviations used when describing instructions 198
10.16 Data manipulation instructions 199
10.17 Data transfer instructions 205
10.18 Program manipulation instructions 210
10.19 Status management and other instructions 212
10.20 Index register instructions 213
Problems 213

11 Programming and Applications **214**

11.1 An introduction to programming 214
11.2 8-bit addition 215
11.3 16-bit addition 216
11.4 8-bit subtraction 217
11.5 Bit testing 217
11.6 8-bit x 8-bit multiplication 220
11.7 Flashing light sequence 223
11.8 Switch-controlled flashing light sequence 226
11.9 Simple traffic light control program 227
11.10 Digital-to-analog convertors (DACs) 234
11.11 Waveform generation using a DAC 236

11.12 Double buffering 238
11.13 A simple analog-to-digital convertor (ADC) 240
11.14 A software-driven successive approximation ADC 242
11.15 Interfacing an 8-bit ADC chip to a microprocessor 246
Problems 248

Further reading 250

Solutions 251

Index 252

Foreword

Microprocessors and microcomputers have had far-reaching effects on education, commerce and industry. The technology associated with them has influenced every section of the community. As is often the case with a rapidly developing technology, early textbooks on the subject were written either for the narrow technical specialist or for those with a broad general interest in the subject. This book provides sound coverage of both the hardware and software of microcomputers, since a knowledge of both is vital to those wishing to understand microcomputer operation. The book should be of value to students following degree and TEC courses in electronics, computing, physics and related topics. It will also be of value to others dealing with microprocessors and microprocessor-based systems.

Rather than invent a microprocessor for use in the book, I have chosen to use one of the most popular family of microprocessors as the basis of the book, namely the 8080 family. A few additions to the architecture of the central processing unit and to the instruction set of this family have been necessary to introduce features which are not normally available in it. These features are introduced to the reader as the book unfolds.

The book begins logically by introducing the reader to the basic requirements of a microprocessor in chapter 1. Chapter 2 deals with binary arithmetic procedures; a knowledge of these processes is necessary to a full understanding of data manipulation by the microprocessor. A wide range of logic devices used in microcomputer systems is covered in chapter 3, including logic gates, three-state logic, flip-flops, integrated circuits, decoders and basic input/output ports. Chapters 4 an 5 outline a microcomputer system together with its basic architecture and the bus system used throughout the microcomputer.

Chapter 6 deals with the important aspects of the memory organisation of the microcomputer. This chapter has sections on ROM (also PROM and EPROM), RAM, memory maps and DMA. The work in chapter 7 covers input/output ports, which are the means by which the central processing unit communicates with the 'outside world'; it includes discussions on accumulator I/O, memory-mapped I/O, on-chip I/O, handshaking, programmable I/O and serial data transmission.

The interrelated topics of subroutines, stacks, interrupts and polling are covered in chapters 8 and 9. Chapter 10 is devoted to a detailed description of the microprocessor instruction set and addressing modes. Finally, in chapter 11, applications are considered, ranging from simple binary addition, to multiplication and a range of advanced applications including traffic light control, digital-to-analog conversion, waveform generation and analog-to-digital conversion. References for further study are provided at the end of the book, and problems (with solutions) are included.

I would like to express my thanks to the individuals and organisations who have materially contributed to the book. Particular thanks are due to Mr S. Rakowski, MSc, Senior Lecturer at the North Staffordshire Polytechnic, who has provided much of the support which has made this book possible. The manufacturers who have contributed to the book include Intel Corporation, Zilog Corporation, Advanced Micro Devices, Quarndon Electronics Ltd, and Hewlett-Packard Ltd. I am also indebted to my wife for her help in the preparation of the typescript, and for her patience and understanding during its writing.

Noel M. Morris

1 An Introduction to Microprocessors

This chapter ranges over some of the topics which are dealt with in detail later in the book. It is hoped that this introductory chapter will remove much of the 'mystery' of microprocessors, which can act as a block to their full understanding.

1.1 Digital Computers

A **digital computer** is simply a calculating machine which carries out a wide variety of operations under the control of a **program of instructions** contained in its **store** or **memory**. The basic blocks associated with a typical computer are illustrated in figure 1.1.

The user communicates with the **central processing unit** (CPU) of the computer by means of peripheral units which contain a range of input devices and output devices. **Input devices** cause **data** to be applied to or to be 'input' to the CPU, and include switches, transducers, sensors, teletypes (a teletype is a form of typewriter which generates an electrical code pattern which the computer can understand), tape recorders, etc. In turn, the computer communicates with the 'outside' world (the user's world) by means of **output devices** which include lamps, light-emitting

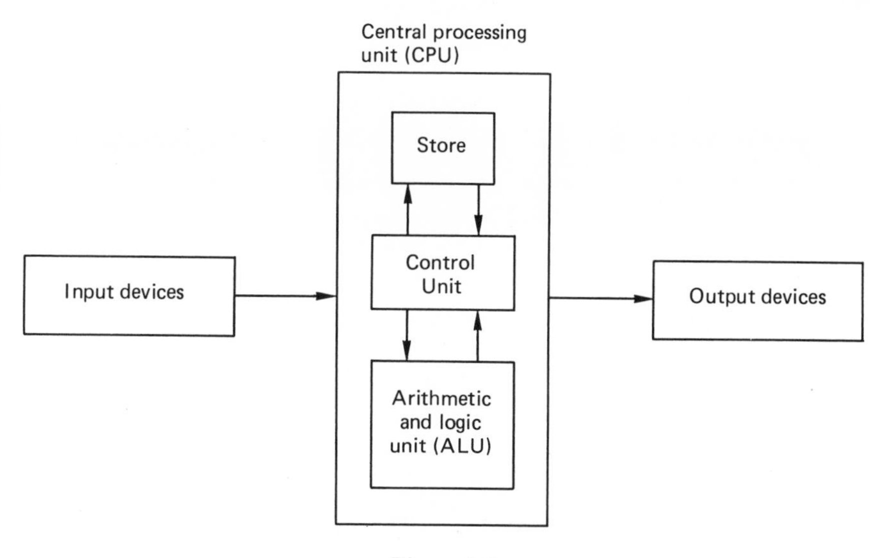

Figure 1.1

diodes, teletypes, graph plotters, printers, video display units, television monitors, etc.

The CPU is the section of the computer which handles data and instructions, and contains the sections described below. The **arithmetic and logic unit** (ALU) is the section in which arithmetical operations (x, ÷, +, -) and logic operations (AND, OR, NOT - see also chapter 3) are carried out on the operands (numbers) in the computer. An important section of the ALU is known as the **accumulator**, which is a working space in the ALU; the accumulator is sometimes known as the **A-register**. The accumulator is used as a temporary storage location for many operations. The ALU also contains several other registers which are accessible to the programmer for use as temporary storage accommodation or for data manipulation.

The **store** or **memory** section is used to store both numbers and instructions that are to be used in the program which the computer executes. A **program** is simply a sequence of instructions written down for the solution of a particular problem. Under the influence of the **control unit**, the instructions are obeyed in sequence so that the program is executed in an orderly manner. For example, when carrying out a MULTIPLICATION operation, the multiplicand (the number being multiplied) must first be transferred from one location in the store and placed in the accumulator under the control of the control unit. The control unit must then cause the multiplier to be taken from a second location in the store. Next the control unit must cause the two numbers to be multiplied together, the product being left in the accumulator (the value of the multiplicand being eliminated in the multiplication process). Finally, under the control of the computer program, the control unit must cause the product in the accumulator to be copied into a third location in the store.

Figure 1.2 (Reproduced by permission of Quarndon Electronics Ltd)

In addition to the above features, the computer needs an accurate timing source to ensure synchronism between the various operations described above. The timing source is usually a crystal-controlled oscillator known as a **clock** source.

Computers are available in all sizes, the largest machines can deal with, for example, the payroll, ordering and invoicing of international organisations. **Microcomputers** are at the other end of the size range, and can consist of a few integrated circuit (IC) 'chips' (see also chapter 3).

A microcomputer consists of a microprocessor (see section 1.3) together with a number of IC chips which allow the user to communicate with it. The prime function of the small system in figure 1.2 may be, for example, as a 'dedicated' controller which is permanently connected to a system such as a set of traffic lights, and is not called on to perform any other function.

By expanding the input/output and memory capabilities of the microcomputer, it becomes a useful general-purpose computer. A typical example of a microprocessor-based general-purpose computer is shown in figure 1.3; this type of microcomputer is portable, yet is capable of handling complex calculations and control problems.

Fitting between the microcomputer and the large computer comes the **minicomputer** which, prior to the introduction of the microcomputer, covered the full range of digital calculations and control problems from very simple problems up to the lower end of the capability of large machines.

Figure 1.3 (Reproduced by permission of Hewlett-Packard Ltd)

1.2 Hardware, Software and Firmware

These terms are widely used to describe features of computers and microprocessors. Digital system **hardware** consists of the components used in its construction; that is to say items of equipment that can physically be touched. **Software** comprises the programs written to allow the computer to execute instructions. Somewhere between hardware and software are instructions which are vital to the starting-up and shut down of the computer as well as to input–output procedures of the system. These are known as **firmware**, and are located in what is known as the **read-only-memory** (ROM) section of the store. These instructions are programmed into hardware chips, and cannot normally be altered by the user.

1.3 A Basic Microcomputer

A basic microcomputer of the type outlined here differs from its larger relative the mainframe computer in several respects, the most important being that it is physically small and much cheaper. A microcomputer system consists of a microprocessor in the form of a semiconductor silicon 'chip' together with a range of support chips. A basic microcomputer system is shown in figure 1.4.

In this case, the microprocessor chip contains the ALU, the control unit and the clock or timing device. The elements contained in the microprocessor chip vary from one microprocessor to another; for example, one microprocessor may contain only the ALU and the control unit, but another type may not only contain the ALU, control unit and clock but also some memory and input/output (I/O) facility.

Microcomputers are organised around a **three-bus** (busbar) system (see figure 1.4) which services all the units. The **control bus** contains several wires and carries out the important functions of informing the system when to 'read' data from memory or from an input device, or when to 'write' data into memory or into an output device, etc. The **address bus** usually contains sixteen lines, and gives the CPU the capability of 'addressing" up to $2^{16} = 65\,536$ individual locations (a 'location' can either be an 'address' in the memory, or it can be an input device or an output device). The number of lines in the **data bus** depends on the type of microprocessor; the majority have eight lines in the data bus, but some use only four lines and others use sixteen. Selected lines from each of the three buses are connected to the various devices along the 3-bus system.

The memory or store is in two sections – the **read-only memory** (ROM) and the **random access memory** (RAM); the latter is more accurately described as a read-write memory since information can either be written into the memory or can be read from it. A point worth noting here is that each location in ROM can also be accessed in a random fashion (that is, any location can be addressed), but it is only possible to read data from each location. The data stored in the ROM is vital to the operation of the microcomputer and must not be lost when the supply is turned off; it is therefore programmed in the form of firmware (see also section 1.2). The RAM may contain both the program that the CPU is operating on and also the relevant

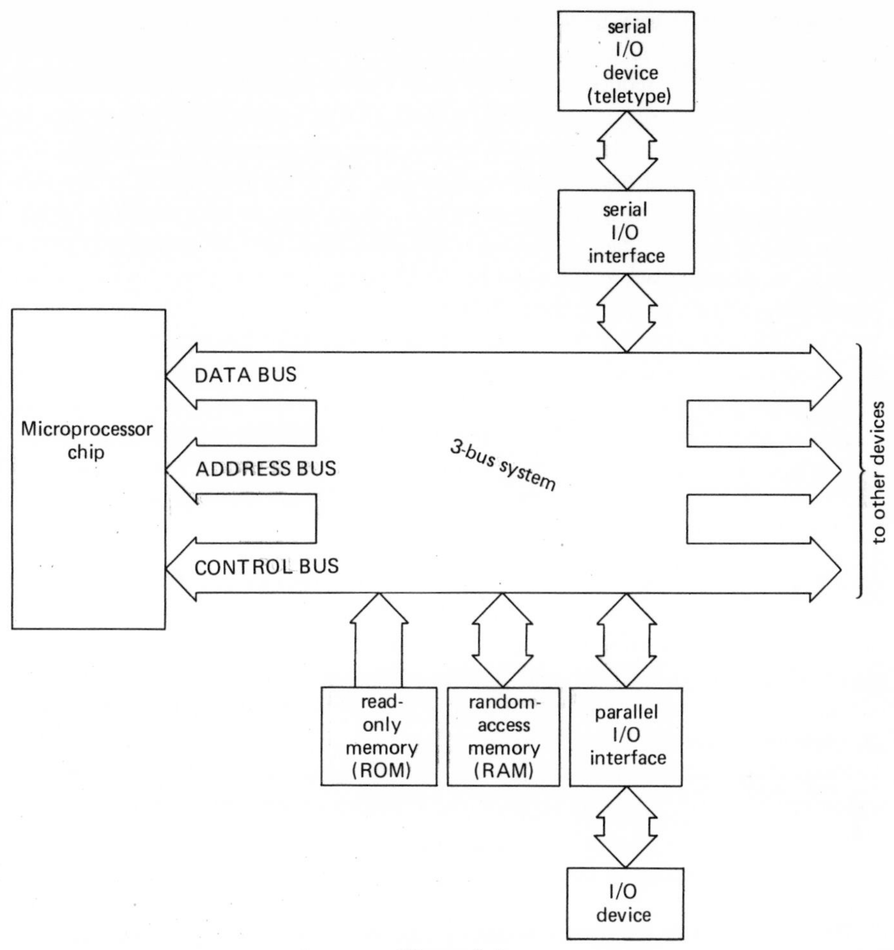

Figure 1.4

data. In most cases the RAM retains its program and data only so long as the power supply to it is maintained. If the power supply fails, the information stored in the RAM will vanish unless preventative action is taken in the few microseconds that are available before the supply voltage falls below the critical operating level. A RAM of this type is described as a **volatile memory**. A ROM is a **non-volatile memory** since it retains its information in the event of a power supply failure. Although the majority of RAMs are volatile, a number of non-volatile types are available to the microcomputer user.

Input/output devices are needed through which we can communicate with the CPU; these are known as **interface units**. The input terminals of the CPU are often known as **input ports** and the output terminals as **output ports**. Broadly speaking there are two types of interface units, namely those which accept a serial input and those which accept a parallel input.

A **serial input** is one in which the data is presented in sequential form. An example of this kind is where the microcomputer is used to count a number of pulses which occur in sequential random order (as, for example, in a traffic census). The CPU itself operates as a 'parallel' system, that is one in which all the data is handled simultaneously. The function of the serial interface unit is to convert the serial data into parallel form. In the case of figure 1.4, the serial I/O interface unit is used in connection with a teletype which causes data to be transmitted every time a key is pressed. Each typewriter character must first be converted from a series of electrical pulses at the input to the I/O interface, into a parallel form at its output before being passed to the CPU.

A **parallel I/O** interface is required where a number of independent lines have signals on them simultaneously. This may occur, for example, where a system has switches and transducers installed at different points which must be 'read' by the computer; alternatively, a parallel output port allows the CPU to send signals simultaneously to a number of lamps, relays or motors.

1.4 Bits, Nibbles and Bytes

An electrical device has one of two operating states, namely either on or off. That is, it is a two-state or **binary** device. If the device provides an output voltage of zero volts when it is OFF and, say, five volts when it is ON, the zero volts level can be described as **logic '0'** and the five volts level as **logic '1'** (these are, alternatively, described as **low** and **high**, respectively). Thus a **binary digit (bit)** may have the logic value of either '1' or '0'.

From the above we can see that a binary variable can have only one of the two conditions '0' or '1'. Applying 'logic' to the situation we see that

$$1 = \text{NOT } 0 = \overline{0}$$

and

$$0 = \text{NOT } 1 = \overline{1}$$

Writing a bar over a **binary variable** (see above) implies that we are dealing with the **logical complement** of the variable (that is, '0' is the logical complement of '1', and vice versa). Complementing a binary variable is also described as logical **negation** or logical **inversion**.

A sequence of four bits such as 1001 or 0010 is sometimes described as a **nibble**. A contiguous (or adjacent) group of eight bits such as 10101010 is known as a **byte**. The number of bits used to convey data in a microprocessor is known as its **word length**; in the majority of microprocessors this is one byte or eight bits, in others it is four bits and in yet others it is sixteen bits. A typical 8-bit word in a microprocessor may be

m.s.b. → 11100011 ← l.s.b.

where the left-hand '1' is described as the **most significant bit** (m.s.b.) and the right-hand '1' is the **least significant bit** (l.s.b.). In the 8080 family of microprocessors the above word could represent the instruction 'compare the accumulator contents with the next number in the program'.

1.5 Input/output Ports

In its simplest form, an input/output port is simply a number of switches which connect the input terminals of the port to the output terminals of the port. Practical ports are more involved than this (see chapter 7), but we shall confine our attention here to a rudimentary type of port. Examples of rudimentary input and output ports are illustrated in figure 1.5a and b, respectively.

The **input port** has eight **data input** lines (designated DI), commencing with line zero (DI_0) and finishing at line seven (DI_7). The reader should note that, in general, the first line is number zero; a few manufacturers give number 'one' to the first line, so that the input lines would be numbered DI_1 to DI_8, but this is not usually the case in microprocessor systems.

The signals applied to the DI lines of the input port are derived from switches or sensors and are either at logic '0' (0 V) or are at logic '1' (say +5 V relative to earth). In the case of the input port in figure 1.5a, the states of eight independent switches can be simultaneously monitored. In the case of the input port in figure 1.5, the switches *inside* the input port (and inside the output port for that matter) are operated electronically by logic signals applied to the **chip select** lines CS1 and CS2; these switches are open-circuit when CS1 and CS2 have logic '0' applied to them. When CS1 has logic '1' applied to it, the left-hand bank of switches closes simultaneously; when CS2 has logic '1' applied to it, the right-hand bank of switches closes simultaneously (in practical I/O ports it is sometimes the case that a '1' causes one of the banks of switches to open, and a '0' causes it to close - this is discussed later in this section). Thus CS1 and CS2 must have logic '1' simultaneously applied to them before the signals on the DI lines are connected to the **data out** (DO) lines and thence to the data bus. How the signals applied to CS1 and CS2 are obtained is discussed in section 1.6.

As mentioned above, the designation CS means chip select; that is when CS1 = 1 and CS2 = 1 simultaneously then the chip has been 'selected'. Alternative designations for these inputs are CE (**chip enable**), DS (**device select**) and DE (**device enable**).

The function of the chip select lines is to ensure that only one port is allowed access to the data lines at any one time; this condition is ensured by the electronics engineer who designs the system hardware.

When the microprocessor wishes to communicate with the outside world it does so through an **output port** (see figure 1.5b). In this case the inputs to the ports are derived from the DATA bus. Once the chip has been selected by means of logic 1's on lines CS1 and CS2, the logic signals (0 V or 5 V) on the DATA bus are connected to the output devices. If the output device on output lines DO_2 and DO_4 of the output port are lamps, then if the signal on DATA line D_2 is logic '1' and that on line D_4 is logic '0', then when the output port is selected the lamp in line DO_2 is illuminated and that in line DO_4 is extinguished.

In the above case we have considered the chip select lines to be 'active' in the 'high' state (the logic '1' state). In the case of many practical chips, one (or even

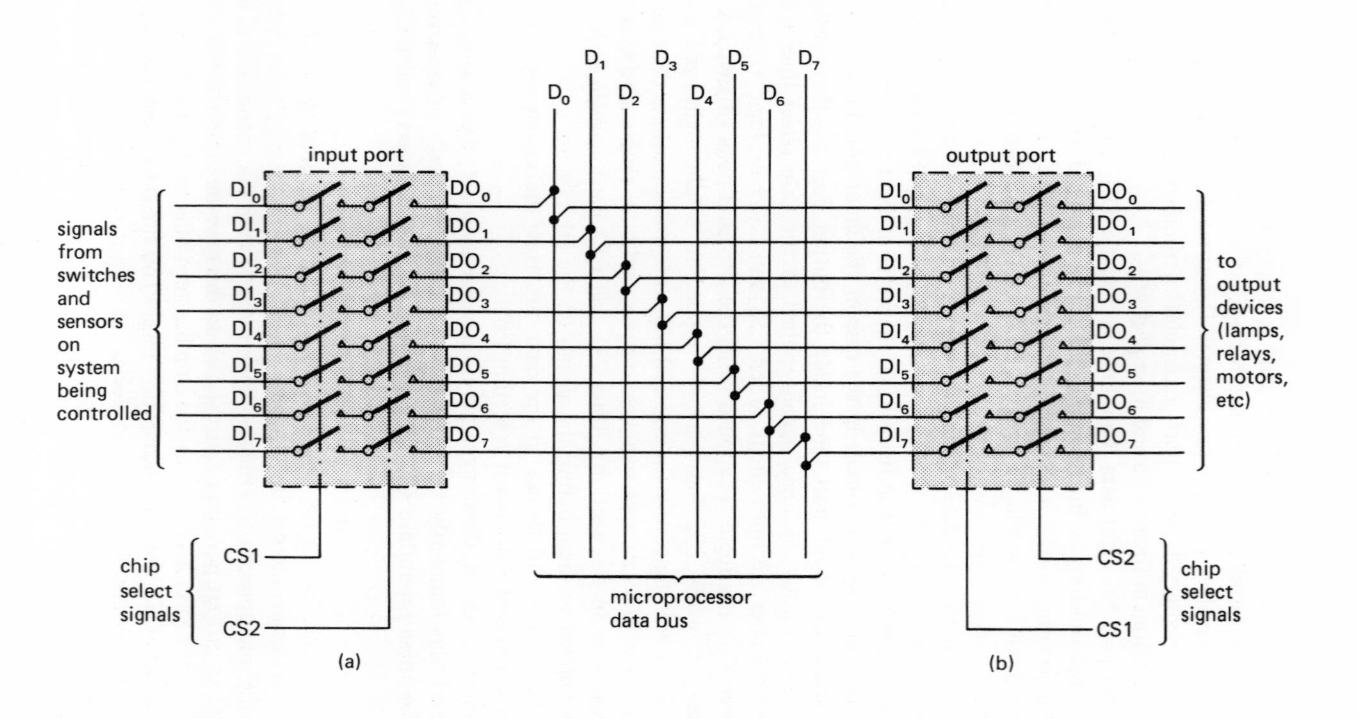
input port
output port
DI_0 DI_1 DI_2 DI_3 DI_4 DI_5 DI_6 DI_7
DO_0 DO_1 DO_2 DO_3 DO_4 DO_5 DO_6 DO_7
D_0 D_1 D_2 D_3 D_4 D_5 D_6 D_7
signals from switches and sensors on system being controlled
to output devices (lamps, relays, motors, etc)
CS1
CS2
chip select signals
microprocessor data bus
(a)
(b)

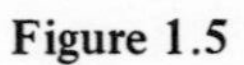
Figure 1.5

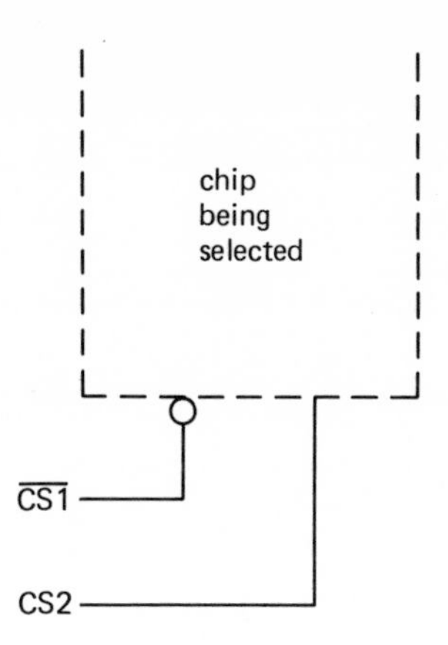

Figure 1.6

both) of the chip select lines are active when the signal applied is logic '0' (that is, they are 'active low'). An example of this kind is illustrated in figure 1.6; here the chip is selected when CS1 is low (logic '0') AND CS2 is high (logic '1') simultaneously. The fact that CS1 is 'active low' is indicated by the circle at the point where the $\overline{CS1}$ (NOT CS1) line connects with the chip; the circle indicates logical inversion.

In some cases I/O ports are **unidirectional**, that is they can transmit information in one direction only; we have assumed in the above discussion that both ports are unidirectional. Certain types of I/O port are **bidirectional**, that is they can transmit information in either direction.

Since the DATA bus not only transmits data from the input port to the CPU, but also from the CPU to the output devices, it is a **bidirectional bus.**

A problem which arises with the simple output port in figure 1.5b, is that the output device will only receive signals during the period of time when the port is 'selected'. Since the port is selected only for a very short period of time (about 0.5 μs), the data is not continuously available at the output terminals. A practical I/O port overcomes this problem by using what are described as 'latches' or 'flip-flops' in the port. The function of these devices is to retain or to memorise the last logical instruction applied to it. For example, if the signal on line D_4 of the DATA bus was '1' at the instant that the output port was selected, then the data latches within the output port ensure that the signal at output line DO_4 of the output port remains at '1' thereafter until new data is applied when the port is next selected. The operation of latches or flip-flops is described in chapter 3.

1.6 A Simple Microprocessor with I/O Ports

A simple microprocessor having two unidirectional input ports (chips A and C) and two unidirectional output ports (chips B and D) is illustrated in figure 1.7. The system uses an 8-bit data word which is transmitted in parallel along the eight wires

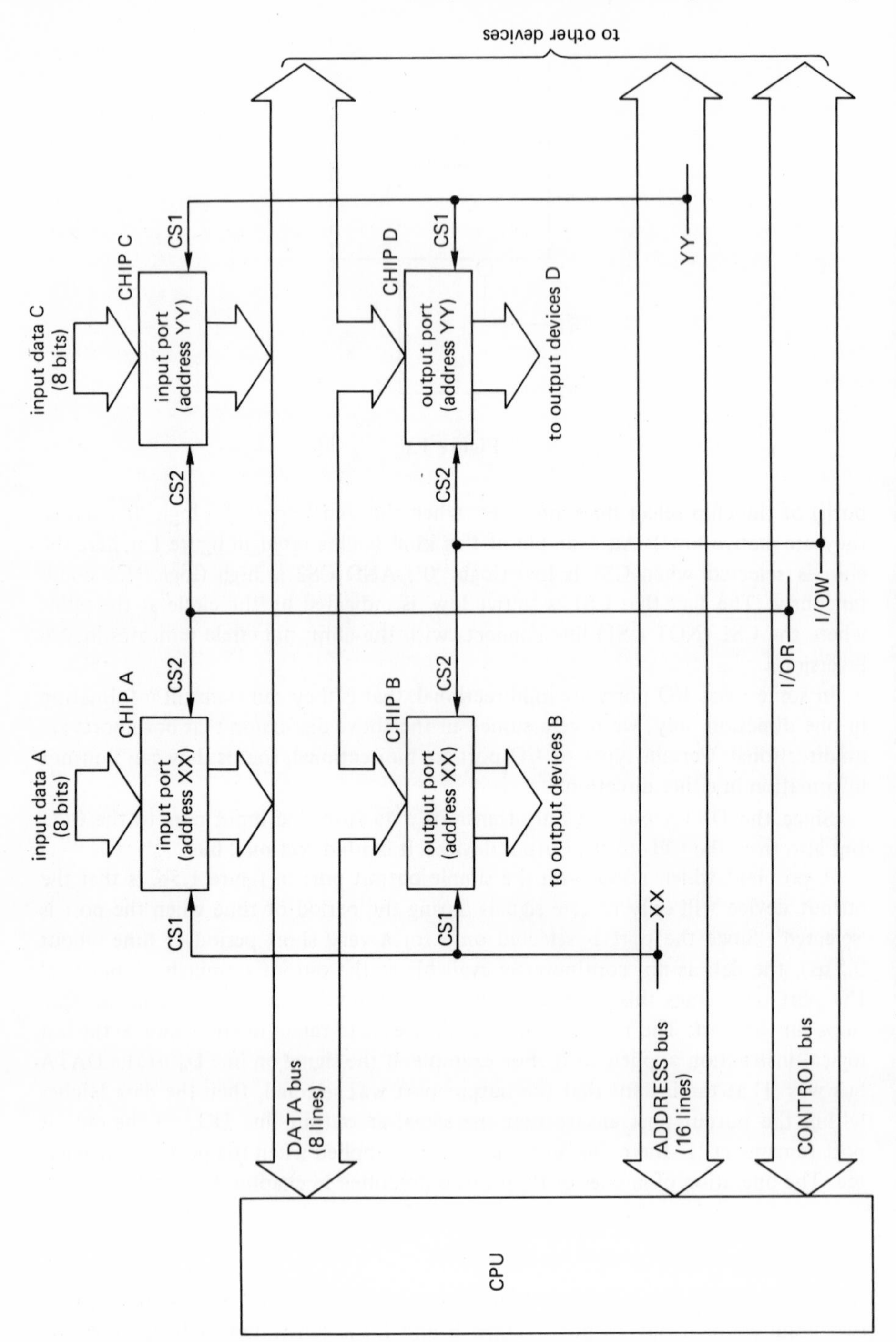
to other devices
input data A
(8 bits)
input port
(address XX)
CHIP A
CS1
CS2
input data C
(8 bits)
input port
(address YY)
CHIP C
CS1
CS2
output port
(address XX)
CHIP B
CS1
CS2
to output devices B
output port
(address YY)
CHIP D
CS1
CS2
to output devices D
XX
YY
I/OR
I/OW
DATA bus
(8 lines)
ADDRESS bus
(16 lines)
CONTROL bus
CPU

Figure 1.7

of the DATA bus, each of the eight wires being connected to the input and output ports in the manner outlined in figure 1.5.

In order to sample data from a particular location, say at the input to CHIP A, the CPU must select CHIP A yet must not select any other chip. As mentioned earlier, this is achieved by means of the chip selection signals CS1 and CS2 associated with each chip. One of the chip select lines, say CS1, is selected by means of a logic signal on one of the lines in the ADDRESS bus (in our case these are arbitrarily described as lines XX and YY. Chip select input CS2 is activated by a logic signal on one of the lines in the CONTROL bus (in our case these are the I/OR [input/output read] and I/OW [input/output write] lines). When the program instructs the CPU to 'read' the state of an input port, it causes a logic '1' to be applied to the I/OR line in the control bus (note: in some CPUs the 'inactive' logic level is logic '1', and the 'active' level is logic '0', so that the 'read' control line would be labelled $\overline{\text{I/OR}}$ and is 'active' when logic '0' is applied to it). When the program instructs the CPU to 'write' data into an output port, it causes a logic '1' to be applied to the I/OW control line (or a '0' if the line is labelled $\overline{\text{I/OW}}$).

The reader will note that CHIP A and CHIP B are given the same 'address', which is XX, corresponding to the line used in the ADDRESS bus. However, only one of the two chips is selected when line XX is activated with a logic '1' signal, since the CS2 input line on each chip is activated by different CONTROL bus signals (the CPU sends out logic 1's at different times on the I/OR and I/OW lines).

Thus CHIP A is selected when ADDRESS XX = 1 AND I/OR = 1 simultaneously; CHIP B is selected when ADDRESS XX = 1 AND I/OW = 1 simultaneously; CHIP C is selected when ADDRESS YY = 1 AND I/OR = 1 simultaneously; CHIP D is selected when ADDRESS YY = 1 AND I/OW = 1 simultaneously. Using this method of addressing I/O ports, it is possible to 'read' data into the CPU from any input port and, if necessary, after processing the data in some way it can be 'written' into any output port.

In section 1.7 a simple application of the microcomputer in figure 1.7 is considered.

1.7 A Simple I/O Sequence

Let us consider the series of events which causes the state of eight switches to control the state of operation of eight independent lamps by means of a microprocessor (see figure 1.8).

The switches are connected to the input terminals of input port A, whose address is 01. The lamps are connected to the output terminals of output port D whose address is 02 (these addresses are selected by the hardware designer).

It is first necessary to **IN**put to the accumulator the contents of address **01** (port A), after which we must **OUT**put the data in the accumulator to address **02** (port D). Each input signal is obtained from a single-pole, double-throw switch which can apply either a logic '0' or a logic '1' to the appropriate input line of port A (see figure 1.8). Each instruction in a program is given a number, and in

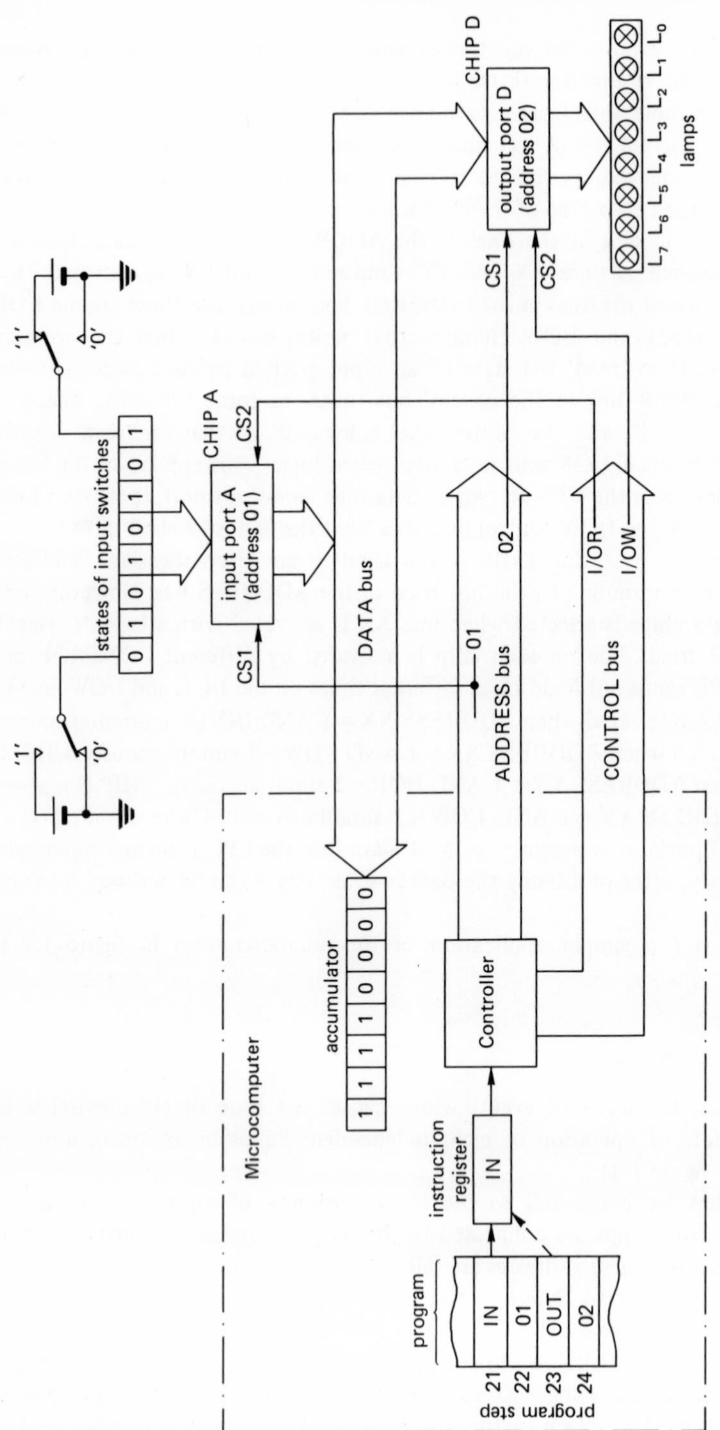
'1'
'0'
states of input switches
0 1 0 1 0 1 0 1
CHIP A
CS1
CS2
input port A
(address 01)
DATA bus
01
ADDRESS bus
02
CONTROL bus
I/OR
I/OW
CHIP D
output port D
(address 02)
CS1
CS2
L_7 L_6 L_5 L_4 L_3 L_2 L_1 L_0
lamps
Microcomputer
accumulator
1 1 1 1 0 0 0 0
Controller
instruction register
IN
program
program step
21 IN
22 01
23 OUT
24 02

Figure 1.8

the following we assume that the first twenty program steps have been executed. As a result of the final instruction acted upon (instruction step 20), the accumulator will contain some information which is irrelevant to the sequence we are concerned with. Consequently, when the CPU carries out the I/O sequence above, it will cause the apparently random data in the accumulator (given as 11110000 in figure 1.8) to be replaced by data fed from the switches connected to port A as follows.

Since we assume that the first twenty instruction steps have been executed, the first instruction in the I/O sequence is located in address 21 in the store. The I/O sequence is written in simple mnemonic form in table 1.1.

Table 1.1 Simple I/O instruction sequence

Program step (or memory address)	Address content	Mnemonic instruction	Comment
21	IN	IN 01	Input data from port 01
22	01		
23	OUT	OUT 02	Output data to port 02
24	02		

Storage location 21 contains the binary equivalent of the INput instruction, which ultimately causes data to be 'input to the accumulator'. The IN instruction is transferred to the **instruction register** (IR) of the microprocessor; this instruction causes the control circuitry to recognise that data are to be read from an input device, and the I/OR (I/O READ) line in the CONTROL bus is activated with a logic '1' signal. This situation is illustrated in figure 1.8. The CPU then inspects the next instruction in the sequence (program step 22), and this results in ADDRESS line 01 being activated. Since both CS1 and CS2 lines of port A are simultaneously activated, the data at the input of port A is transferred to the DATA bus and thence to the accumulator. Thus the data formerly held by the accumulator is replaced by the new data fed in from port A (see table 1.2). The data in the accumulator is therefore changed to 01010101.

Table 1.2

Program step	Instruction	Contents of accumulator	Port A input	Port D output
21	IN	11110000	01010101	00000000
22	01	01010101	01010101	00000000
23	OUT	01010101	01010101	00000000
24	02	01010101	01010101	01010101

Having completed program step 22, the CPU interprets the next byte of data at step 23 as an instruction. The control circuit recognises the contents of step 23 as 'output the contents of the accumulator to the address specified in the next program step' (see the broken arrow in figure 1.8). Accordingly, signals are applied to the I/OW line of the CONTROL bus and line 02 of the ADDRESS bus (the signals on the CONTROL bus line I/OR and ADDRESS bus line 01 meanwhile being removed, so that the inputs to port A are isolated from the DATA bus). This causes the contents of the accumulator (which is now 01010101) to be applied to the outputs of port D (see table 1.2); that is L_0, L_2, L_4 and L_6 are illuminated and L_1, L_3, L_5 and L_7 are extinguished.

The reader should note that after program step 22 has been executed by the CPU, the state of the switches connected to port A can be altered, but any such alteration will not affect the pattern illuminated when the instruction OUT 02 has been executed, that is L_0, L_2, L_4 and L_6 remain illuminated. The reason is that after the IN 01 instruction has been executed, CHIP A is isolated from the DATA bus. Any change in input data is therefore not transmitted to the DATA bus.

1.8 Conclusion

In this chapter the reader has been presented with a simple introduction into the way in which data is moved between input and output ports, the CPU acting as an intermediate device in the process. Before discussing microprocessor 'architecture' in detail, it is necessary for the reader to understand something about binary arithmetic and also logic circuits. These topics are dealt with in the following two chapters.

PROBLEMS

1.1 If the address of the I/O chips in figure 1.7 are

port XX - address 01
port YY - address 02

write a program which results in the logical condition of the sensors connected to input port YY being transferred to the output devices connected to output port XX.

1.2 Draw up a table for problem 1.1 which shows the movement of data between the ports (see, for example, table 1.2).

1.3 For a system having I/O ports with the addresses in problem 1.1, write a program which causes the logic state of the sensors connected to input port YY being transferred to output port XX and also to output port YY. The program must then cause the state of the sensors connected to input port XX to be transferred to output port XX.

2 *Binary Numbers and Arithmetic*

2.1 Numbering Systems

A wide range of numbering systems is in use today, the most common being the **decimal** or **denary system.** This system utilises the ten digits 0, 1, 2, 3, 4, 5, 6, 7, 8 and 9. It is important to note that the **first** number is zero, and that the **tenth** number is nine. The *total* number of digits used in the system is known as the **base** or **radix** of the system; the radix of the decimal system is ten. A numbering system widely used in association with computers is the **octal system** which has a radix of eight and uses the eight characters 0, 1, 2, 3, 4, 5, 6, 7. Yet another system which is widely used in association with microprocessors is the **hexadecimal system** (abbreviated to **hex**) having a radix of sixteen.

Let us consider the structure of numbers. Consider the denary number 431.6; this number is given by

$$431.6_{10} = (4 \times 10^2) + (3 \times 10^1) + (1 \times 10^0) + (6 \times 10^{-1})$$

In order to identify the radix of the number, it is written as a subscript (as in the case of the above example). Each digit is **scaled** by the position of the decimal point in the number; this 'point' gives the position where the power to which the radix is raised changes sign from positive to negative. Thus the whole number part is to the left of the decimal point, and the fractional part of the number is to the right of the decimal point. In general, we shall refer to the **radix point,** although in specific systems we give it a specific name such as the decimal point in the decimal system and the binary point in the binary system.

The above form of number representation is used in all numbering systems as illustrated below

$$\text{hexadecimal } 98.2 = 98.2_{16} = (9 \times 16^1) + (8 \times 16^0) + (2 \times 16^{-1})$$
$$\text{octal } 10.5 = 10.5_8 = (1 \times 8^1) + (0 \times 8^0) + (5 \times 8^{-1})$$
$$\text{binary } 10.1 = 10.1_2 = (1 \times 2^1) + (0 \times 2^0) + (1 \times 2^{-1})$$

The reader should note that whilst we refer to the number 10_{10} as 'ten', the number 10_2 is described as 'binary one, zero'. The reason for this difference is that 10_{10} corresponds to ten units, whereas 10_2 corresponds to $(1 \times 2^1) = 2_{10}$ units

The numbers above are written down in what is known as **fixed-point notation.**

That is the radix point is fixed at the point where the power of the radix changes from positive to negative. However, when we carry out a complex calculation such as multiplication, it is necessary to keep track of the radix point so that it can be correctly positioned in the final solution. Whilst this is a relatively simple operation for humans, it is a time consuming chore for electronic systems. Consequently, many digital systems use **floating-point notation** for arithmetic operations; this helps to minimise the 'housekeeping' associated with keeping track of the radix point. Number representation in the floating point format is illustrated below. The number 205.6_{10} can be represented in any of the following forms

$$205.6_{10} = 205.6 \times 10^0 = 20.56 \times 10^1 = 2.056 \times 10^2 = .2056 \times 10^3$$

Every time we shift the radix point to the left, the power of the multiplying radix factor is increased by unity. Thus three left shifts causes the radix multiplying factor to be raised to three. Binary numbers can be represented in this way, as illustrated below

$$1011.01_2 = 1011.01 \times 2^0 = .101101 \times 2^4 \qquad (2.1)$$

When the binary point has been shifted to the left of the m.s.b. of the binary number, it is said to be in its **normalised** floating-point form. When a digital system operates in floating-point notation, the numbers are stored in normalised form. In general, a floating point number X is represented in floating point form as follows

$$X = m \times r^e$$

where m is the **mantissa** (or **fractional part** or **argument**) of the number (which is 101101 in expression 2.1), r is the **radix** of the system (which is 2 in expression 2.1), and e is the **exponent** (or **characteristic**) of the number (which is 4 in expression 2.1).

A point which must be borne in mind when dealing with number systems in computers is that the 'length' of the number is restricted by the word length of the computer. In the majority of microprocessors the word length is eight bits; thus if we wish to store the integer 101_2 in a microprocessor with a one-byte word length, it must be stored in the form 00000101, where the five most significant zeros are known as **non-significant zeros.**

The first twenty-four numbers in a selection of systems are listed in table 2.1; non-significant zeros have not been shown. The reader will note that when all the digits in a system have been used once, the value in the units column becomes zero and a '1' appears in the next higher column. In the decimal system this occurs after 9_{10} (which is the tenth digit, counting zero as the first digit). In the octal system it occurs after 7_8 (which is the eighth number in the system); when all the characters have been used up once more in the units column (at 17_8), the next number is 20_8. Since the binary system has a radix of two, both values have been used after 1_2 is reached, and the next value is 10_2 (or 2_{10}); the value following this is 11_2 (which corresponds to $2_{10} + 1_{10} = 3_{10}$), etc.

The reader will observe from table 2.1 that N binary digits allow us to define up to 2^N decimal values. Thus a two bit ($N = 2$) combination enables us to define up to

Table 2.1

	System			
	Binary	Octal	Decimal	Hexadecimal
Radix	2	8	10	16
	0	0	0	0
	1	1	1	1
	10	2	2	2
	11	3	3	3
	100	4	4	4
	101	5	5	5
	110	6	6	6
	111	7	7	7
	1000	10	8	8
	1001	11	9	9
	1010	12	10	A
	1011	13	11	B
	1100	14	12	C
	1101	15	13	D
	1110	16	14	E
	1111	17	15	F
	10000	20	16	10
	10001	21	17	11
	10010	22	18	12
	10011	23	19	13
	10100	24	20	14
	10101	25	21	15
	10110	26	22	16
	10111	27	23	17
	11000	30	24	18

$2^2 = 4$ values (these are the values 00, 01, 10 and 11); in a four bit combination ($N = 4$) there are $2^4 = 16$ groups of bits ranging from 0000_2 (0_{10}) to 1111_2 (15_{10}).

The hexadecimal system has a radix of sixteen. Since humans are familiar with the decimal system, we can use the ten digits of the decimal system (0–9) to represent the first ten digits of the hex system. Beyond this point we must devise a new set of characters to represent the remaining numbers in the hex system; that is we need characters to represent the equivalent of 10_{10} to 15_{10}. It has become the practice to use the first six letters of the alphabet for this purpose. Thus $A_{16} = 10_{10}$, $B_{16} = 11_{10}$, ... $E_{16} = 14_{10}$ and $F_{16} = 15_{10}$. The number in the hex system following F_{16} is 10_{16}, etc. Thus the hex number FAB_{16} is given by

$$FAB_{16} = (15 \times 16^2) + (10 \times 16^1) + (11 \times 16^0) = 4011_{10}$$

The hex numbering system has become popular with microprocessor users because one hex character is equivalent to four binary digits. For example

$$F_{16} = 1111_2$$
$$A_{16} = 1010_2$$
$$9_{16} = 1001_2$$

Since most microprocessors have a word length of eight bits, then a complete binary word can be expressed by two hex characters.

2.2 Converting a Number of any Radix into its Decimal Equivalent

A number N can be expressed in the form of a power series or polynomial as follows

$$N = d_n r^n + d_{n-1} r^{n-1} + \ldots + d_1 r^1 + d_0 r^0 + d_{-1} r^{-1} + \ldots + d_{-n} r^{-n}$$

where d is an integer in the range zero to $(r-1)$, r is the radix of the system and n is an integer. For example, the decimal number 180.2 can be represented by the series

$$180.2_{10} = (1 \times 10^2) + (8 \times 10^1) + (0 \times 10^0) + (2 \times 10^{-1})$$

where $N = 180.2, d_2 = 1, n = 2, d_1 = 8, n-1 = 1, d_0 = 0$ and $d_{-1} = 2$. **To convert any non-decimal number into its decimal equivalent, the number is expanded as a polynomial in powers of the radix, the decimal value being given by the sum of the individual terms in the series** as follows

$$\begin{aligned}
110.1_2 &= (1 \times 2^2) + (1 \times 2^1) + (0 \times 2^0) + (1 \times 2^{-1}) \\
&= (4 + 2 + 0 + 0.5)_{10} = 6.5_{10} \\
67.3_8 &= (6 \times 8^1) + (7 \times 8^0) + (3 \times 8^{-1}) \\
&= (48 + 7 + 0.375)_{10} = 55.375_{10} \\
AFC.B_{16} &= (A_{16} \times 16^2) + (F_{16} \times 16^1) + (C_{16} \times 16^0) + (B_{16} \times 16^{-1}) \\
&= (10_{10} \times 16^2) + (15_{10} \times 16^1) + (12_{10} \times 16^0) + (11_{10} \times 16^{-1}) \\
&= (2560 + 240 + 12 + 0.6875)_{10} = 2812.6875_{10}
\end{aligned}$$

2.3 Converting a Decimal Number into its Equivalent in any Radix

Firstly, the number must be divided into two parts, namely its integral or whole number part and its fractional part. The parts to the left and to the right of the decimal point being dealt with in different ways. We will illustrate the technique by converting 34.625_{10} into its binary form. In this case the integral part of the number is 34 and its fractional part is .625.

Integral part

The rule is to **divide the integer repeatedly by the radix, successive remainders giving the required value** (the first remainder being the least significant digit of the resulting integer).

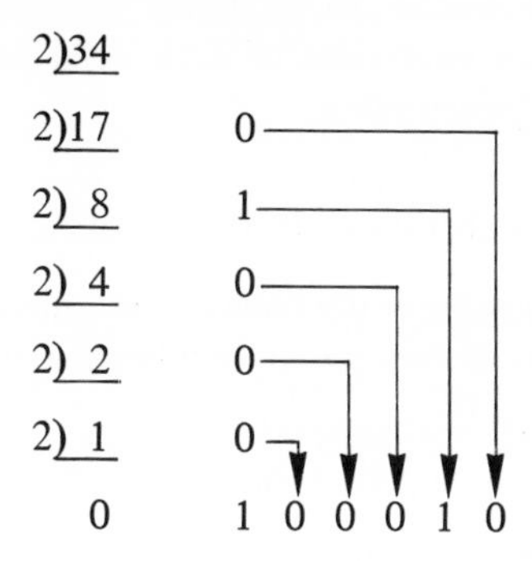

Thus $34_{10} = 100010_2$

Fractional part

The rule is to **multiply the fractional part repeatedly by the radix, the integral part of the result giving successive values in the number** (the first integral part being the most significant digit of the fractional part).

In the example considered the fractional part is .625, and the procedure is illustrated below.

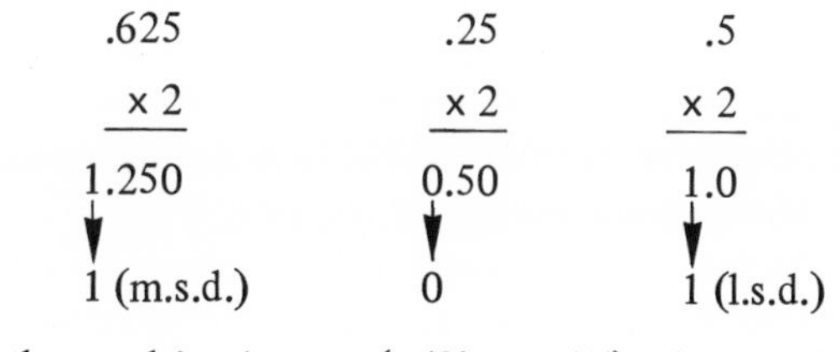

.625
x 2
1.250
.25
x 2
0.50
.5
x 2
1.0
Integral part
1 (m.s.d.)
0
1 (l.s.d.)

After each multiplication, the resulting integer (a '0' or a '1') takes its place in the binary value, the fractional part of the result being carried over to be multiplied once more by the radix. Thus

$$.625_{10} = .101_2$$

Result

Combining the above values gives

$$34.625_{10} = 100010.101_2$$

2.4 Converting a Binary Number into an Octal Number

The rule is as follows: **commencing at the binary point, divide the binary number into groups of three bits, the value of each group being written down in its octal form.** For example

$$111000101.010011_2 = 111\ 000\ 101.010\ 011_2 = 705.23_8$$

When converting an octal number to its binary equivalent, the above process is reversed. That is the binary equivalent of each octal value is written down, the combined result giving the binary value.

2.5 Converting a Binary Number into a Hexadecimal Number

Proceed as follows: **commencing at the binary point, divide the binary number into groups of four bits, the value of each group then being written down in its hex form.** For example

$$111000101.010011_2 = 0001\ 1100\ 0101.0100\ 1100_2 = 1C5.4C_{16}$$

A hex number is converted into its binary equivalent by reversing the process. That is the binary equivalent of each hex digit is written down, the combined result being the binary value.

2.6 Binary-decimal codes

Data is handled in microprocessor systems in binary form. One method of encoding them is in the **pure binary form** in table 2.1. Humans find the decimal system a convenient notation when dealing with numbers, and in order to communicate numerical data between man and machine it is necessary to devise suitable **coding** methods.

A decimal system is one which has ten unique digits 0, 1, . . . 8, 9. From the work in section 2.1 we see that four binary digits can be used to represent $2^4 = 16$ combinations ranging from 0000 to 1111. A decimal system can be devised which is coded in binary form simply by choosing *any ten* of the sixteen possible combinations. Such a system is described as a **binary-coded decimal (BCD) numbering system.** If each digit in the system is given a decimal value or **weight**, the system can be described in terms of the weights of the bits. A number of BCD codes are given in table 2.2.

Consider for a moment the 8421 BCD code. The m.s.b. of the code is given a weight of 8_{10}, the next less significant bit has a weight of 4_{10}, the next lower bit has a weight of 2_{10}, and the l.s.b. has a weight of 1_{10}. This code allows us to uniquely define any decimal number simply by converting each binary '1' to its decimal value as follows

$$(0101)_{8421\ \mathrm{BCD}} = (0 \times 8) + (1 \times 4) + (0 \times 2) + (1 \times 1) = 5_{10}$$

$$(1001)_{8421\ \mathrm{BCD}} = (1 \times 8) + (0 \times 4) + (0 \times 2) + (1 \times 1) = 9_{10}$$

Table 2.2 BCD codes

Decimal value	BCD code			
	Weight	8421	2421	642(−3)
0		0000	0000	000 0
1		0001	0001	010 1
2		0010	0010	001 0
3		0011	0011	100 1
4		0100	0100	010 0
5		0101	1011	101 1
6		0110	1100	011 0
7		0111	1101	110 1
8		1000	1110	101 0
9		1001	1111	111 1

Comparing the 8421 BCD code in table 2.2 with the sixteen groups of four bits in the pure binary code in table 2.1, we see that the groups 1010, 1011, 1100, 1101, 1110 and 1111 are not used. That is to say, they are **forbidden groups** or **illegal groups** in the 8421 BCD code.

When referring generally to BCD codes, reference to 'the BCD code' usually implies that we are talking about the 8421 BCD code rather than any other BCD code (see below).

Many other decimal weights can be used to define BCD codes, the 2421 BCD code in table 2.2 being an example. In this code the m.s.b. has a weight of 2_{10} and the l.s.b. has a weight of 1_{10}. Hence

$$(1011)_{2421\ \text{BCD}} = (1 \times 2) + (0 \times 4) + (1 \times 2) + (1 \times 1) = 5_{10}$$

$$(1111)_{2421\ \text{BCD}} = (1 \times 2) + (1 \times 4) + (1 \times 2) + (1 \times 1) = 9_{10}$$

The reader will note that in the 2421 BCD code it is possible for more than one code group to represent a given decimal value. For example, 5_{10} can be represented as 1011 (see above) or as 0101; 6_{10} can be represented as 1100 (see table 2.2) or as 0110. However, once ten binary groups have been selected to represent ten digits, then the remaining six are forbidden groups; for the 2421 BCD code in table 2.2 the combinations 0101 and 0110 are amongst those which are forbidden.

Codes having negative weights can also be used, the 642(−3) BCD code in table 2.2 being an example. In this case

$$(1011)_{642(-3)\ \text{BCD}} = (1 \times 6) + (0 \times 4) + (1 \times 2) + (1 \times [-3])$$

$$= 5_{10}$$

In general there is no 'best' code for all applications, since the advantages of a particular code depend on many factors including the speed with which it can be used and the complexity of the circuitry needed to encode it and to decode it into, say, decimal. The most popular BCD code is the 8421 BCD code, since it is well understood and the circuitry required is usually available in IC form.

2.7 Multi-number Representation in BCD

Suppose that we wish to represent the decimal value 96_{10} as an 8421 BCD value. In this case the weights of the bits representing the '90' part of the number are each increased by a factor of 10; that is they have the weights 80, 40, 20 and 10, respectively. Thus

$$\begin{aligned} 96_{10} &= [(1 \times 80) + (0 \times 40) + (0 \times 20) + (1 \times 10)] \\ &\quad + [(0 \times 8) + (1 \times 4) + (1 \times 2) + (0 \times 1)] \\ &= (10010110)_{8421\ \mathrm{BCD}} \end{aligned}$$

The reader will note that two decimal digits are represented by eight bits, which is the usual word length in most microprocessors. A fractional number can be dealt with as follows

$$79.36_{10} = (01111001.00110110)_{8421\ \mathrm{BCD}}$$

Decimal addition using BCD values is described later in this chapter.

2.8 Error Detection in Binary Code Groups

When data is transmitted between electronic equipment, there are many possibilities for errors to arise in the received data. An error is said to occur when the transmitted and received data is not equivalent. Possible causes include induced voltage, faulty equipment, etc. To increase the reliability of data transmission, **redundant data** is included in the transmitted binary word.

Redundancy can be illustrated by means of the the English language as follows. If we include a number op errers in thes sintince, the English language is sufficiently redundant to allow us to not only detect but also to mentally correct the errors.

The simplest method of error detection in digital systems is to include an additional bit (known as a **parity bit**) in each code group, the parity bit being redundant so far as data transmission is concerned. There are two simple types of parity check bit used, known as odd parity and even parity, respectively.

If **odd parity** is used, the parity bit (P_1 in table 2.3) is zero if the sum of the 1's in the data part of the word is odd. It is '1' if the sum of the 1's in the data part of the word is even.

If **even parity** is used, the parity bit (P_2 in table 2.3) is zero if the sum of the 1's in the data part of the word is even. It is '1' if the sum of the 1's in the data part of the word is odd.

Suppose that a digital system using odd parity transmits the decimal value 9_{10}

Table 2.3 Odd and even parity check on 4-bit data words

Decimal value	Odd parity P_1	8421	Even parity P_2	8421
0	1	0000	0	0000
1	0	0001	1	0001
2	0	0010	1	0010
3	1	0011	0	0011
4	0	0100	1	0100
5	1	0101	0	0101
6	1	0110	0	0110
7	0	0111	1	0111
8	0	1000	1	1000
9	1	1001	0	1001

using the 8421 BCD code in the data group; the transmitted word is 11001, the m.s.b. being the parity bit. If, during the transmission process a fault occurs and the received data is 11011, then the parity checking network at the receiving end will detect that an error has occurred. Although a simple check of this kind can detect one error, it cannot correct the error. Several parity bits must be used in each word transmitted if an error is to be corrected, for it is not only necessary to detect that an error has occurred, but also the position of the error in the word must be known.

The condition or status of special binary digits such as the parity bit are stored in a microprocessor in a **status word**. The state of the parity bit can be checked by inspecting the state of the status word. The status word is stored in what is known as the **status register** or **flag register**; a 'flag' is a name used to describe an indicator (a binary digit) which gives the result of certain operations (such as evaluating the parity bit).

2.9 Alphanumeric Codes

In order to handle not only numbers but also alphabetical characters and other special characters (for example, symbols such as + (/ . , etc), the microprocessor must be capable of handling a code which allows for at least 26 capital (upper case) letters, 26 lower case letters, 10 numeric values (0–9), and at least 25–30 special characters.

If only thirty special characters are used, the code used to represent the above combination must have 92 combinations. A 6-bit code has $2^6 = 64$ combinations which is insufficient for our purposes. A 7-bit code has $2^7 = 128$ combinations, which is adequate for the majority of alphanumeric work.

Since an 8-bit word length is used in the majority of microprocessors, seven bits can be used for the data code and the eighth bit can be used for parity checking.

Two popular alphanumeric codes are the 7-bit **ASCII code** (American Standard

Code for Information Interface) and the 8-bit **EBCDIC code** (Extended Binary Coded Decimal Interchange Code). The ASCII code can be used by all microprocessors, and a description is given below.

The format of an ASCII code word is shown below, and comprises two groups of four bits; the data associated with the ASCII word is contained in the lower seven bits, that is bits b_0 to b_6, the m.s.b. being retained for parity check purposes.

$$\overbrace{b_7 b_6 b_5 b_4}^{\text{most significant character}} \underbrace{b_3 b_2 b_1 b_0}_{\text{least significant character}}$$

The ASCII code is summarised in table 2.4. The first thirty-two code groups (00_{16}-21_{16} or 00000000_2-00100001_2) and the final group ($7F_{16}$ or 01111111_2) are used for non-printing characters each of whose functions are described in table 2.4. The remaining characters being printing characters including upper case and lower case alphabetical characters, decimal numbers and a range of special characters.

2.10 Binary Addition

Where $x + y = z$, x is the **augend,** y the **addend** and z the **sum**. When the sum of two numbers is greater than the radix of the numbering system, a **carry** is generated. This carry must be added to the next higher column of the calculation (this occurs, for example, in the decimal sum 9 + 1 = zero + a carry of 1). The carry bit is said to be **carried in** to a sum if it is the carry generated by the next lower sum. The carry is said to be **carried out** if it is generated by the addition process under consideration. Thus the carry-out bit of one sum becomes the carry-in bit of the next higher sum.

The possible combinations which exist when adding two binary variables are

Augend	0	0	1	1
Addend	0	1	0	1
Sum	0	1	1	0
Carry-out	0	0	0	1

Also, when adding two binary variables together with a carry-in, the following possibilities occur

Carry-in	1	1	1	1
Augend	0	0	1	1
Addend	0	1	0	1
Sum	1	0	0	1
Carry-out	0	1	1	1

Table 2.4 The ASCII code

Least significant character (hex)

Most significant character (hex)	0	1	2	3	4	5	6	7	8	9	A	B	C	D	E	F
0	NUL	SOH	STX	ETX	EOT	ENQ	ACK	BEL	BS	HT	LF	VT	FF	CR	SO	SI
1	DLE	DC1	DC2	DC3	DC4	NAK	SYN	ETB	CAN	EM	SUB	ESC	FS	GS	RS	US
2	SP	!	"	#	$	%	&	'	(	)	*	+	,	-	.	/
3	0	1	2	3	4	5	6	7	8	9	:	;	<	=	>	?
4	@	A	B	C	D	E	F	G	H	I	J	K	L	M	N	O
5	P	Q	R	S	T	U	V	W	X	Y	Z	[	\	]	^(↑)	-(←)
6	`	a	b	c	d	e	f	g	h	i	j	k	l	m	n	o
7	p	q	r	s	t	u	v	w	x	y	z	{	|	}	~	DEL

NUL	All zero
SOH	Start of heading
STX	Start of text
ETX	End of text
EOT	End of transmission
ENQ	Enquiry
ACK	Acknowledge
BEL	Audible signal
BS	Backspace
HT	Horizontal tabulation
LF	Line feed
VT	Vertical tabulation
FF	Form feed
CR	Carriage return
SO	Shift out
SI	Shift in
DLE	Data link escape
DC1,DC2,DC3,DC4	Device controls
NAK	Negative acknowledge
SYN	Synchronous idle
ETB	End of transmission block
CAN	Cancel
EM	End of medium
SUB	Substitute
ESC	Escape
FS	File separator
GS	Group separator
RS	Record separator
US	Unit separator
SP	Space
DEL	Delete

We will consider the addition of the two 8-bit numbers 10111011_2 and 10100011_2.

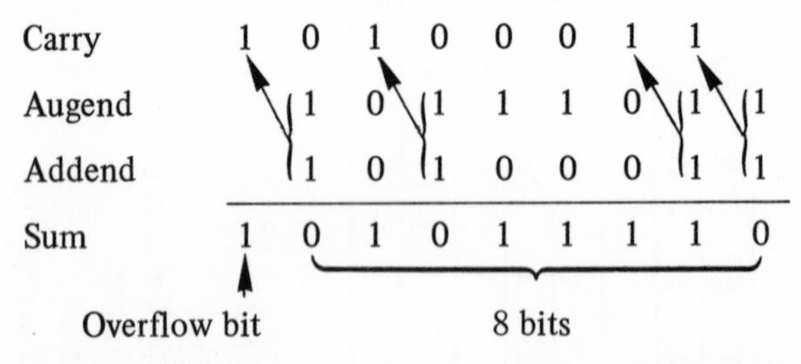

The first column of the addition gives 1 + 1 = 0 + a carry of '1'. This carry is added to the next higher column to give a sum of 1 + 1 + 1 = 1 + a carry of '1'. The reader is asked to follow the addition procedure to its conclusion, when it will be seen that a carry of '1' is generated by the sum of the m.s.b. of the augend and the m.s.b. of the addend. This carry (an **overflow**) gives the m.s.b. of the result, which is 101011110_2. The above example illustrates one limitation of the word length of a microprocessor (or of any digital system for that matter), in that unless care is taken, any overflow may be lost giving rise to a serious error in the result.

2.11 Hexadecimal Addition

The general procedure in hex addition is generally the same as for any other radix, that is a carry-out is generated when the sum of two numbers exceeds the radix of the system (16 in hex). A summary of the addition of two hex numbers is given in table 2.5. From this table we see that

$$5_{16} + 2_{16} = 7_{16}$$

and

$$A_{16} + B_{16} = 15_{16}$$

The addition of the two hex numbers $7A9_{16}$ and $B18_{16}$ is illustrated below

Carry	1	0	1	
Augend		7	A	9
Addend		B	1	8
Sum	1	2	C	1

hence

$$7A9_{16} + B18_{16} = 12C1_{16}$$

2.12 Binary Complement Notation

In the expression $x - y = z$, x is the **minuend**, y the **subtrahend**, and z the **difference**. The concept of a 'minus' sign is man-made and it is necessary to invent some means

Table 2.5

Augend	Addend														
	1	2	3	4	5	6	7	8	9	A	B	C	D	E	F
1	2	3	4	5	6	7	8	9	A	B	C	D	E	F	10
2	3	4	5	6	7	8	9	A	B	C	D	E	F	10	11
3	4	5	6	7	8	9	A	B	C	D	E	F	10	11	12
4	5	6	7	8	9	A	B	C	D	E	F	10	11	12	13
5	6	7	8	9	A	B	C	D	E	F	10	11	12	13	14
6	7	8	9	A	B	C	D	E	F	10	11	12	13	14	15
7	8	9	A	B	C	D	E	F	10	11	12	13	14	15	16
8	9	A	B	C	D	E	F	10	11	12	13	14	15	16	17
9	A	B	C	D	E	F	10	11	12	13	14	15	16	17	18
A	B	C	D	E	F	10	11	12	13	14	15	16	17	18	19
B	C	D	E	F	10	11	12	13	14	15	16	17	18	19	1A
C	D	E	F	10	11	12	13	14	15	16	17	18	19	1A	1B
D	E	F	10	11	12	13	14	15	16	17	18	19	1A	1B	1C
E	F	10	11	12	13	14	15	16	17	18	19	1A	1B	1C	1D
F	10	11	12	13	14	15	16	17	18	19	1A	1B	1C	1D	1E

of telling the digital system that it is handling a negative quantity. One method of dealing with this situation is illustrated below.

Suppose that we are dealing with a binary system which has a word length of five bits and is capable of handling numbers, but is incapable of displaying '+' and '–' signs. The way in which the number –1 is stored in the machine is illustrated by subtracting unity from zero as follows

$$0_2 = (0)0000$$

$$1_2 = \underline{(0)0001}$$

$$\text{SUBTRACT} \qquad (1)1111 \equiv -1_2$$

Thus -1_2 can be represented in a 5-bit binary system by (1)1111. In the above case the m.s.b. of the word is known as the **sign bit**, and for our convenience is enclosed in parentheses; when stored in a microprocessor, the sign bit is identified by the fact that it occupies the m.s.b. of the number. Hence, in the notation adopted here, **positive numbers have a sign bit of zero, and negative numbers have a sign bit of unity**. Moreover, the magnitude of a positive number is given by the value of the four least significant bits (the data bits); in the above example this is 0001_2 or $+1_2$. The magnitude of a negative number is represented in one of its **complement** forms (in the above case, the number -1_2 is held in what is known as the 2's complement form [described below]). Thus (1)1111 is the 2's complement representation of -1_2.

In the binary system there are two types of complement representation of negative numbers - the **2's complement** (also known as the **true complement** or the **radix complement**) and the **1's complement** (also known as the **reduced radix complement**). Both are described below.

1's complement representation

To obtain the 1's complement representation of a binary number the rule is **beginning with the l.s.b., change the 0's to 1's and 1's to 0's.** Examples of the 1's complement of several 6-bit binary numbers are given in table 2.6, the m.s.b. being the sign bit.

Table 2.6 Binary complements

	Binary number	1's complement	2's complement
(a)	(0)000.00	(1)111.11	(0)000.00
(b)	(1)000.00	(0)111.11	(1)000.00
(c)	(0)101.10	(1)010.01	(1)010.10
(d)	(1)101.10	(0)010.01	(0)010.10
(e)	(0)011.11	(1)100.00	(1)100.01
(f)	(1)011.11	(0)100.00	(0)100.01

It is important to note that non-significant zeros (both less than the least significant '1' and greater than the most significant '1') must also be complemented.

Consider the number in row (c) of table 2.6; the binary value considered in the left-hand column of table 2.6 is $+101.10_2$ (or $+5.5_{10}$). The negative equivalent of this value (-101.10_2 or -5.5_{10}) is represented in the 1's complement form as (1)010.01. In row (d) of table 2.6 we see from the sign bit (1) that the number in the left hand column is a negative number; moreover, this (1) tells us that the number is stored in the 1's complement notation. Thus (1)101.10 is equivalent to -010.01_2 (that is, -2.25_{10}); when we evaluate the 'negative' equivalent of number (d) the result will be a positive number since $-(-010.01_2) = +010.01_2$

2's complement representation

To obtain the 2's complement representation of a binary number, carry out one of the following rules.

(1) **Beginning with the l.s.b., change the 0's to 1's and 1's to 0's (that is, form the 1's complement) and then add '1' to the l.s.b. of the number so formed.**

(2) **Commencing with the l.s.b., copy the number up to and including the least significant '1', all bits thereafter being 1's complemented (that is, 0's changed to 1's and 1's to 0's).**

The reader is invited to apply both rules to verify the 2's complement values in table 2.6.

2.13 Binary Subtraction

Two numbers can be subtracted from one another as follows

$$x - y = x + (-y)$$

That is we can 'add' the negative equivalent of the subtrahend (the number being subtracted) to the minuend (the number being subtracted from). In the case of the binary system, subtraction can be carried out by adding the 2's complement of the subtrahend to the minuend (the 1's complement can be used, but the procedure is not as straightforward as using the 2's complement - see, for example, *Logic Circuits* by Noel M. Morris (McGraw-Hill, London)). The general rule for subtraction when using the signed 2's complement notation is as follows.

The 2's complement of the subtrahend is added to the minuend. If the sign bit of the result is '0', then the result is the true difference and is given a positive sign. If the sign bit is '1', the result is the 2's complement of the difference; the true difference is given by the 2's complement of the result, and is given a negative sign. Any overflow or carry produced by the calculation is 'lost'.

Two examples of the use of the above rule using an 8-bit word length are given below

(a) (0)0011010 - (0)0001101 (or $26_{10} - 13_{10}$)

The 2's complement of (0)0001101 is (1)1110011, hence

Minuend	26_{10}	≡ (0)0011010
Subtrahend	-13_{10}	≡ (1)1110011
ADD		1(0)0001101
Overflow (lost) ↗		

Since the sign bit is '0' then the result is the true difference, that is (0)0011010 - (0)0001101 = $(0)0001101_2 = 13_{10}$.

(b) (0)0001101 - (0)0011010 (or $13_{10} - 26_{10}$)

The 2's complement of (0)0011010 is (1)1100110, hence

$$\begin{array}{lrl} \text{Minuend} & 13_{10} & \equiv (0)0001101 \\ \text{Subtrahend} & -26_{10} & \equiv \underline{(1)1100110} \\ \text{ADD} & & (1)1110011 \end{array}$$

Since the sign bit is '1', the result is in 2's complement form. That is (0)0001101 − (0)0011010 = (1)1110011 = $-(0)0001101_2 = -13_{10}$

2.14 Hexadecimal Subtraction

As mentioned earlier, the hexadecimal notation is well suited to microprocessors since an 8-bit word can be represented by two hex characters.

Hexadecimal subtraction can be performed by complement addition, much as for binary subtraction. In this case we need to use the **16's complement** (the radix complement) of the subtrahend; this is determined as follows.

Subtract each hex character from F_{16}, after which 1_{16} is added to the least significant hex character of the result.

As an example we will determine the 16's complement of $06A_{16}$, where the most significant character is the sign character. Firstly, the number is subtracted from $(F)FF_{16}$ to give $(F)95_{16}$. Next, 1_{16} is added to the least significant character to give

$$-6A_{16} = (F)96_{16}$$

The sign character (F) implies that the number is a negative value and is represented in the 16's complement form.

Example 2.1 Subtract $2A6_{16}$ from $B1_{16}$ (or $177_{10} - 678_{10}$)

Solution Since $+2A6 = (0)2A6_{16}$, then

$$\begin{aligned} -2A6_{16} &= [(F)FFF - (0)2A6] + 1 = (F)D59 + 1 \\ &= (F)D5A_{16} \end{aligned}$$

Before we can subtract the two numbers, it is necessary to adjust the length of the minuend so that it is equal to that of the subtrahend as follows

$$\begin{array}{rl} +B1_{16} & = (0)0B1 \\ -2A6_{16} & = \underline{(F)D5A} \\ \text{ADD} & (F)EOB \end{array}$$

Since the sign character is (F), the number is negative and is represented in 16's complement form. Hence

$$B1_{16} - 2A6_{16} = (F)EOB_{16} = -[(0)1F4_{16} + 1_{16}] = -1F5_{16} = -501_{10}$$

2.15 Multiword Binary Arithmetic

With the usual word length of 8 bits we can deal with numbers in the range 0–255 (corresponding to 2^8 combinations); if one of the bits is used for sign purposes, then we can deal only with numbers in the range 0–127 (corresponding to 2^7 combinations). Moreover, if the eight bits are used to store decimal number in BCD format, then only 100 values can be handled in the range 0–99 (this arises from the fact that eight bits can handle only two 4-bit BCD values). In order to improve the accuracy of calculations in microprocessors, it may be necessary to use several computer words to represent a given number. For example the decimal number −13729 can be represented by two 8-bit words as follows

```
                high byte  low byte
sign bit ——► (1)1001010 01011111
             \_____ 16-bit number _____/
```

It is important to remember that all eight bits of the least significant word (the **low byte**) are used for data, but in a signed number the m.s.b. of the most significant word (the **high byte**) is used as the sign bit.

If the length of the binary word is 8 bits, then 8-bit mathematical operations are said to result in a **single precision** result. When a result is evaluated to 16 bits, it is said to be a **double precision** result. Some microprocessors have a double-length addition in their instruction set (in the 8080/8085 family the mnemonic for this is DAD). Examples of 16 bit addition and subtraction are given below; the reader will note that the calculations are completed in two steps, the low byte calculation is completed first and any carry from the low byte is added to the l.s.b. of the high byte.

Example 2.2 Add 2002_{16} to $327A_{16}$ (or $8194_{10} + 12\,922_{10}$)

Solution

Augend	(0)0110010	01111010
Addend	(0)0100000	00000010
ADD	(0)1010010	01111100
Sign bit ↗	5 2	7 C_{16}

hence

$$327A_{16} + 2002_{16} = 527C_{16} = 21\,116_{10}$$

Example 2.3 Using 2's complement notation, subtract $327A_{16}$ from 2002_{16} (or $8194_{10} - 12\,922_{10}$)

Solution The magnitude of the subtrahend is (0)0110010 01111010, hence its 2's complement is (1)1001101 10000110.

Minuend	(0)0100000	00000010
Subtrahend	(1)1001101	10000110
ADD	(1)1101101	10001000

Since the sign bit is '1', the result is in 2's complement form, hence

$$2002_{16} - 327A_{16} = -\underbrace{0010}_{1}\underbrace{010}_{2}\ \underbrace{0111}_{7}\underbrace{1000}_{8_{16}} = -1278_{16} = -4728_{10}$$

2.16 BCD (decimal) Addition

Since man understands the decimal system, it would be helpful in some applications to arrange for the microprocessor to handle decimal numbers directly. Since a decimal version of the binary system exists (the 8421 BCD code), it only remains to devise suitable procedures to handle decimal arithmetic using the BCD code. In the following we discuss BCD addition.

The 8421 BCD code (simply referred to as the BCD code in this section) uses four bits to encode the decimal numbers 0-9. However, the combinations $1010_2 - 1111_2$ ($A_{16} - F_{16}$) are illegal or disallowed, illustrated in figure 2.1.

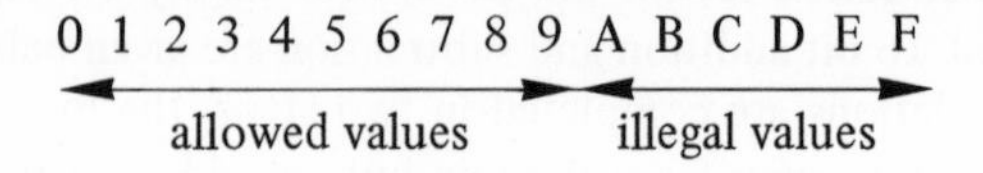

Figure 2.1

Consequently we must arrange whenever necessary for the decimal (BCD) arithmetic process to 'jump' over the disallowed states. A rule which allows the microprocessor to do this is given below.

If the sum of the numbers either (1) generates a forbidden code (that is, A_{16} to F_{16}) or (2) generates a carry, add 6 to the resulting sum. If neither of the above occurs, the result is correct.

The above process is known as **decimal adjustment**. The reader should note that in the case where the result is correct, the 'decimal adjustment' is zero; consequently, zero is added to the result of the first addition (see example 2.4a below).

In some microprocessors a 'decimal add' instruction is included in the instruction set (defined by the mnemonic DAD in the INS 8060 SC/MP instruction set); in others a pure binary addition is first carried out, followed by the decimal adjustment procedure described above (which is given by the mnemonic DAA in the instruction set of the 8080/8085 family).

Example 2.4 Using BCD addition determine the result of the following:

(a) 2 + 6, (b) 4 + 8, (c) 9 + 8.

Solution

	(a)	(b)	(c)
Augend	2	4	9
Addend	6	8	8
ADD	8	C	1
Carry	0	0	1
Decimal adjustment	00	06	06
Result	08	12	17

2.17 BCD Addition in Microprocessors

As mentioned earlier, microprocessors either have a decimal (BCD) addition instruction or use some form of decimal adjustment procedure following a binary (or hex) addition.

A microprocessor having a word length of one byte can deal with two decimal (BCD) numbers, the most significant decimal value being defined by the four m.s.b.s in the 8-bit word, and the least significant decimal value by the four l.s.b.s. In the following we will consider the decimal adjustment procedure used in the 8080 family. In decimal (BCD) operations the 8-bit word is divided as follows

$$\text{m.s.b.} \rightarrow \underbrace{b_7 b_6 b_5 b_4}_{\text{high group}} \underbrace{b_3 b_2 b_1 b_0}_{\text{low group}} \leftarrow \text{l.s.b.}$$

When we add two decimal (BCD) numbers together (each in the range 00 to 99), the CPU adds the corresponding 8-bit binary words in pure binary, and then proceeds to make a decimal adjustment to the low group and the high group as follows (remember, after the addition the sum is in the accumulator).

(1) **If, after the addition, the decimal equivalent of the least significant four bits of the accumulator is greater than 9 (that is, value in the range A_{16} to F_{16}) or if the auxiliary carry (A_C) flag (see below) is set to '1', the binary equivalent of 6 is added to the four l.s.b.s. of the accumulator.**

(2) **If, after the addition, the decimal equivalent of the most significant four bits of the accumulator is greater than 9 (that is, a value in the range A_{16} to F_{16}) or if the carry (C) flag is set to '1', then the binary equivalent of 6 is added to the four m.s.b.s of the accumulator.**

The 'flag' register has been mentioned earlier; one of the flags is used to indicate the state of the auxiliary carry (that is, the carry from the four l.s.b.s) and another is used to indicate the state of the final carry. The A_C flag is set to '1' if a carry bit is transferred between bits b_3 and b_4 of the accumulator during the addition operation. The C flag is set to '1' if a carry occurs from bit b_7 during the addition operation.

Example 2.5 Add the following decimal values, assuming that they are stored in BCD form in a microprocessor: (a) 24 + 13, (b) 78 + 24, (c) 46 + 70, (d) 87 + 99.

Solution The reader will note that in the following, the addition is carried out in two steps. First the augend and the addend are added together in hex (the reader may find it an interesting exercise to perform the addition in binary); this is described as the '1st ADD' operation. The state of the C and A_C flags are then investigated to give the appropriate 'Decimal Adjustment to Accumulator' (DAA), that is, either add 6 or add 0. This is described as the '2nd ADD' operation.

	(a)	(b)	(c)	(d)
Augend	24	78	46	87
Addend	13	24	70	99
C, A_C	0 0	0 0	0 0	1 1
Result of 1st ADD	37	9C	B6	20
DAA	00	06	60	66
C	0	1	1	0
Result of 2nd ADD	37	02	16	86
Result	$03\,7_{10}$	$10\,2_{10}$	$11\,6_{10}$	$18\,6_{10}$

An advantage of performing the above addition in hex is that it is obvious when an illegal code group is produced by an addition. It is not quite so obvious when a 4-bit binary group is considered.

To illustrate the mechanism of the addition process, consider solution (d). The sum of 7 and 9 gives 10_{16} (or 10000_2); that is a sum of zero, and the A_C flag is set to '1'. This '1' is carried forward and added to (8 + 9), giving $1 + 8 + 9 = 12_{16}$ or 10010_2. Since both the C and A_C flags are set to '1', then 6 must be added to each value in the sum, that is, 66 is added to the result of the first addition. The net result in the 10^0 column is $0 + 6 = 6$, in the 10^1 column is $2 + 6 = 8$, and in the 10^2 column is $1 + 0 = 1$. The final result of (d) is therefore 186_{10}.

2.18 Binary Multiplication

If $x \times y = z$, x is the **multiplicand**, y the **multiplier** and z the **product**. Multiplication is a form of repeated addition; for example, the calculation 386 x 34 can be performed in one of several ways, viz:

(a) Add 386 to itself thirty-three times.
(b) Add 386 to itself three times, shift one place to the left and add 386 three times to the sum.

The process of multiplying two binary numbers by the shifting and adding process is illustrated by multiplying the binary equivalent of 7_{10} by 5_{10}.

```
Multiplicand        111
Multiplier          101
                    ---
                    111
                   000
                  111
                  -----
Product           100011   (35₁₀)
```

The reader will note from the above that the product of two n-bit numbers requires up to $2n$ bits to accommodate the result.

If one of the two numbers to be multiplied together is negative, the magnitude of the product can be evaluated by multiplying the magnitude of the two values (this involves taking the 2's complement of the negative value); since one of the numbers is negative, the result must finally be converted into its complement form.

In a digital system, the process of shifting and adding is mechanised as illustrated below. If A (the multiplicand) and B (the multiplier) are multiplied together in a microprocessor, the value of A is stored in a register which we will call the **multiplicand register**, and B is stored in another register known as the **multiplier register**; the product is stored in a *double-length* register known as the **product register**, P. Register A retains the same data throughout the multiplication process. The contents of register B are shifted right at each stage of the calculation; each 'shift right' of the contents of register B causes the l.s.b. of B to vanish, the m.s.b. being filled with a '0'.

When multiplying two binary numbers each of length *n* bits, the process is complete after *n* add and *n* shift-right steps have been carried out.

The l.s.b. of the multiplier (B) is known as the **multiplication bit**, M. The n most significant data bits (not including the overflow bit) of the product register are described as P_M in the following. The multiplication sequence begins with an ADD and is completed after the nth right-shift. The steps are outlined below.

ADD: If M = 0, add zero to P_M (see table 2.7)
If M = 1, add the contents of register A to P_M

SHIFT: Shift the contents of register B and the complete contents of register P one place to the right.

When multiplying two binary numbers together, it is necessary either to allow for an overflow bit to accommodate any overflow which may occur during the calculation (as it does in the example below), or it is necessary to restrict the 'length' of the numbers so that an overflow cannot occur. Thus, if two 3-bit numbers are to be multiplied together, the total length of the product register must either be 7 bits (the m.s.b. being used as temporary accommodation for any overflow) or, alternatively, if the register can only accommodate 6 bits then the value of the product must be 31_{10} or less (that is, equal to or less than 011111_2 to ensure that an overflow does not occur).

Example 2.6 Multiply 111_2 by 101_2 (that is, $7_{10} \times 5_{10}$) using the method outlined above.

Solution The complete process is illustrated in table 2.7. The result after the third SHIFT and ADD sequence is

$$111_2 \times 101_2 = 100011_2 = 35_{10}$$

Table 2.7

Multiplicand register	Multiplier register, B	Double-length product register, P	
111	101	(0) $\underbrace{000}_{P_M}$ 000	1st ADD and SHIFT
	↑	↑ Overflow bit	
		M = 1; add A to P_M	
	M	(0) 111 000	
	↓	Shift B and P one place right	
	010	(0) 011 100	2nd ADD and SHIFT
		M = 0; add 0 to P_M	
		(0) 011 100	
		Shift B and P one place right	
	001	(0) 001 110	3rd ADD and SHIFT
		M = 1; add A to P_M	
		(1) 000 110	
		Shift B and P one place right	
	000	(0) 100 011 = RESULT	

2.19 Binary Division

If $x/y = z$, then x is the **dividend**, y the **divisor** and z the **quotient**. The simplest method of dividing one number by another is by a process of repeated subtraction and shifting. Consider $1001_2 \div 11_2$

```
     11
 11)1001
    11
    ──
     11        hence 1001₂ ÷ 11₂ = 11₂
     11
    ──
     00
```

The process of division can be mechanised in a number of ways. A popular method is known as the **non-restoring method of division**; an algorithm for this method is given below.

(1) **Subtract the divisor from the dividend by 2's complement addition.**
(2) **Note the value of the sign bit of the difference and proceed as follows.**
   (a) **Sign bit = 0; record a '1' in the partial quotient register; shift the difference and the partial quotient one place to the left; subtract the divisor from the difference.**
   (b) **Sign bit = 1; record a '0' in the partial quotient register; shift the difference and the partial quotient one place to the left; add the divisor to the difference.**

As an example of the application of non-restoring division, consider $1111_2 \div 11_2$ (or $15_{10} \div 3_{10}$). The application of the above algorithm is shown in table 2.8. The first operation is to subtract the divisor from the dividend by adding its complement of (1)01. The reader should note that the m.s.b.s of the data part of the divisor and dividend are aligned in the subtraction process; since the subtraction is carried out by complement addition, the m.s.b. of the data part of the divisor is '0'. Since the remainder from this operation is positive (indicated by a sign bit of '0'), a '1' is recorded in the partial quotient register.

The contents of the partial quotient register and that of the dividend register are then shifted one place left, the sign bit of the dividend register being replaced in this operation by the m.s.b. of the data part of the dividend register. Since the remainder is positive, the divisor is subtracted once more from the contents of the dividend register.

The result of this subtraction gives a sign bit of '1', so that '0' is recorded in the partial quotient register; the contents of the dividend register and the partial quotient register are then shifted left once more.

The shift and add (or subtract) process is repeated until the desired accuracy is obtained. If the result is needed to N places, the process is repeated N times. The position of the binary point is evaluated as follows. If the order or power of the dividend and the divisor are p and q, respectively, then the power of the most

Table 2.8

Non-restoring division			
Divisor	Dividend	Partial quotient	Comment
(0)11	(0)1111	.....	
↑ sign bit →	(1)01		Subtract divisor
carry (lost) →	1(0)0011	1	Sign bit '0'; record '1'
	(0)011	...1 .	Shift left
	(1)01		Subtract divisor (rule 2(a))
	(1)101	...10	Sign bit '1'; record '0'
	(1)01	..10 .	Shift left
	(0)11		Add divisor (rule 2(b))
carry (lost) →	1(0)00	..101	Sign bit '0'; record '1'
	(0)0	.101 .	Shift left
	(1)01		Subtract divisor (rule 2(a))
	(1)01	.1010	Sign bit '1'; record '0'
	(0)1	1010 .	Shift left
	(0)11		Add divisor (rule 2(b))
	(1)01	10100	Sign bit '1'; record '0'

significant '1' in the quotient is 2^{p-q}. In the above example $p = 3$ and $q = 1$, hence the power of the most significant '1' in the quotient is $2^{3-1} = 2^2$. Therefore $1111_2 \div 11_2 = 101.00\ldots$

PROBLEMS

2.1 Convert the following numbers into hexadecimal: (a) 01101111_2, (b) 11001110_2, (c) 79_{10}, (d) 94_{10}.

2.2 Devise two BCD codes (other than those in table 2.2), one having a positive weighting and one a negative weighting.

2.3 Represent the number 596_{10} in the two codes devised in problem 2.2.

2.4 Add together the binary numbers 110110110 and 100001111. Express the answer in hexadecimal.

2.5 Add together the hexadecimal numbers 7F and F7. Express the result (a) in hex, (b) in octal.

2.6 Write down the 1's complement value and the 2's complement value of the following binary numbers (the m.s.b. is the sign bit): (a) 10000000, (b) 00000001, (c) 10101010, (d) 01010101.

2.7 Convert 38_{10} and 15_{10} into their 8-bit binary equivalents and, using binary notation throughout, determine the result of (a) $38_{10} - 15_{10}$, (b) $15_{10} - 38_{10}$. Represent the result in binary.

2.8 Repeat problem 2.7 but using the hexadecimal notation.

2.9 Using binary throughout, multiply the numbers 1010_2 and 101_2.

2.10 Using binary throughout, divide 1010_2 by 101_2.

3 *Logic Devices*

3.1 Truth Tables

As described earlier, every problem in the world of logical algebra has either a 'true' solution or a 'false' solution. The 'true' and 'false' states are presented in terms of logic signals, that is, as electrical voltages. The 'true' condition is usually defined as a logic '1' signal and the 'false' condition as logic '0'. The operation of any logic circuit can therefore be described by means of a table containing 1's and 0's which define the state of the input signals and output signals. Such a table is known as a **truth table**.

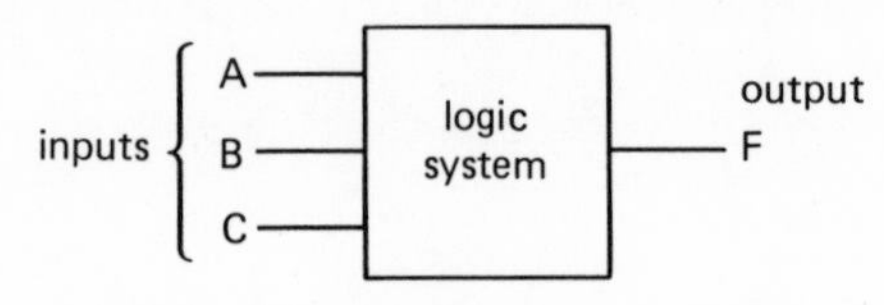

Figure 3.1

A simple logic system is depicted in figure 3.1. Suppose that the figure represents an alarm system having input signals A, B and C which are derived from sensors detecting respectively smoke, a broken window and forced entry through a door. The output signal, F, is used to activate the alarm signal. Let us suppose that the operation of the system is described in the truth table in table 3.1.

Table 3.1 Truth table of a logic system

Inputs			Output
A	B	C	F
0	0	0	0
0	0	1	1
0	1	0	1
0	1	1	1
1	0	0	1
1	0	1	0
1	1	0	1
1	1	1	1

Under 'healthy' conditions, the output from each sensor is logic '0' (which indicates a normal operating condition). When smoke is detected or a break-in occurs, the appropriate sensor produces a logic '1' at its output terminal, that is, A = 1 when smoke is detected. The output F, from the system may either be logic '0' or logic '1', depending on the combined state of sensors A, B and C. The truth table provides a specified set of output conditions which satisfy a particular customer's requirements; if another customer has a different set of requirements corresponding to the same set of sensors, it is only necessary to alter the 1's and 0's in the output column to satisfy the new requirements.

The attention of the reader is now directed to table 3.1, where it will be seen that the output signal F is logic '0' either when A = 0 AND B = 0 AND C = 0 simultaneously (first row of the table) OR when A = 1 AND B = 0 AND C = 1 simultaneously (sixth row). Under all other conditions the output is logic '1', and the alarm must be activated. As mentioned above, the 1's and 0's in the output column of the truth table depend on the requirements of the system, which may differ from those considered here. Consider for the moment the sixth row of table 3.1. Here A = 1 corresponds to smoke being detected and C = 1 corresponds to forced entry to the premises; the truth table indicates that it is a requirement of the system that the alarm is silent (F = 0) when A = 1 AND B = 0 AND C = 1 (corresponding to smoke and forced entry); however unlikely this specification, a logic circuit can be designed to satisfy it.

If a logic system deals with N **variables** or input signals, there are 2^N combinations of these variables in the input column of the truth table. The reason for the above is that each input signal can have any one of two possible values, namely logic '1' or logic '0'. Table 3.1 deals with three variables A, B and C, so that the table has $2^3 = 8$ rows, each row corresponding to one of the eight combinations of A, B, and C.

If the system handles only two variables, then the truth table has $2^2 = 4$ rows, and if it has four input variables it has $2^4 = 16$ rows.

The 1's and 0's in the output column depend on the specification of the logic network, which is often defined by the user of the system (that is, it is *user defined*).

3.2 Logic Signal Levels

Digital electronic circuits have specific bands of voltage which correspond respectively to logic '1' and logic '0'. For example, the logic '1' signal from a TTL (transistor–transistor logic) gate may have any voltage in the range 2.5–4.5 V, while the logic '0' level corresponds to a voltage in the range 0.1–0.4 V. Another type of gate using CMOS (complimentary metal-oxide-semiconductor) logic with the same value of supply voltage may give a logic '1' output voltage of practically 5 V, and a logic '0' output of practically 0 V.

Positive logic notation

In this notation, the electrical potential corresponding to a logic '1' signal is more positive than that associated with a logic '0' signal. For example, if a logic system has two possible output voltages of 3.0 V and 0.25 V, then in the positive logic notation logic '1' is represented by 3.0 V and logic '0' is represented by 0.25 V.

Negative logic notation

In this notation the potential corresponding to a logic '1' signal is less positive than that associated with a logic '0' signal. Using the above voltage levels with a negative logic notation, then logic '1' corresponds to 0.25 V and logic '0' corresponds to 3.0 V.

3.3 Logic Gates

A logic gate is an electronic circuit which can either be 'opened' to or 'closed' to the flow of signals, that is, it 'gates' or controls the flow of data.

The basic range of gates used in electronics are known by the names AND, OR, NOT, NOR and NAND; the name given to a gate describes the logical operation carried out. These gates are described below.

3.4 The AND Gate

Circuit symbols for a two-input AND gate are shown in figure 3.2 and its truth table is given in table 3.2. The reader will note from the truth table that the output, F, is logic '1' only when both **A AND B** are logic '1' simultaneously. The AND

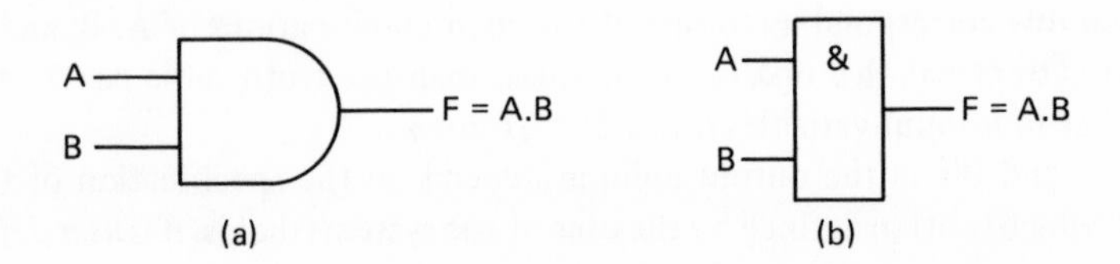

Figure 3.2

Table 3.2 Truth table of a two-input AND gate

Inputs A	Inputs B	Output F = A AND B = A.B
0	0	0
0	1	0
1	0	0
1	1	1

function is symbolised in engineering literature by a dot (.) symbol. The truth table can be summarised by the following.

> **The output from an AND gate is logic '0' whenever any input is logic '0'. Otherwise the output is logic '1'.**

One parameter which indicates the usefulness of a gate is the number of input lines available to the user; this parameter is known as the **fan-in** of the gate. In the case of the gate in figure 3.2, the fan-in is two. Figure 3.3 shows an AND gate with a fan-in of five; that is, it can handle signals from five separate sources. The logical expression for the output from figure 3.3 is

$$F = A \text{ AND } B \text{ AND } C \text{ AND } D \text{ AND } E = A.B.C.D.E$$

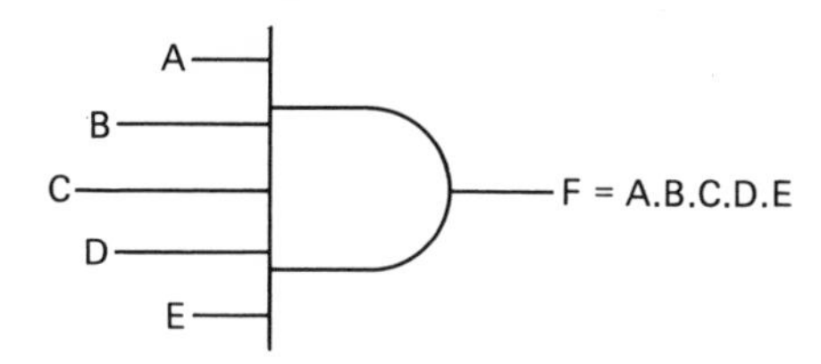

Figure 3.3

The truth table for the 5-input AND gate (not given here) has $2^5 = 32$ rows (that is, there are 32 combinations of 1's and 0's in the input column). There is only one '1' in the output column, which occurs when A, B, C, D AND E are logic '1' simultaneously. The reader will find it an interesting exercise to draw up this truth table.

3.5 The OR Gate

Circuit symbols for a two-input OR gate are shown in figure 3.4, and its truth table is in table 3.3. The reader will note that in the case of the OR gate, the output is

Table 3.3 Truth table for a two-input OR gate

Inputs		Output
A	B	F = A OR B = A + B
0	0	0
0	1	1
1	0	1
1	1	1

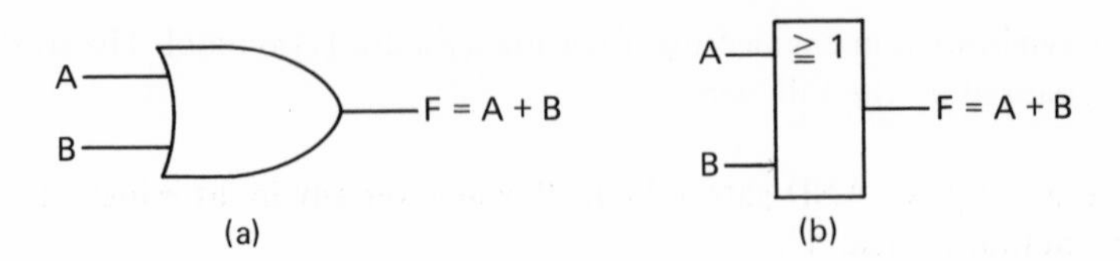

Figure 3.4

logic '1' when either **A OR B** is logic '1'. The OR function is symbolised by a 'plus' (+) symbol; this function should not be confused with the mathematical 'plus' sign since, from the final row of the truth table

'1' OR '1' = '1'

that is, logically

$$1 + 1 = 1$$

The truth table can be summarised as follows:

The output from an OR gate is logic '1' whenever any input is logic '1'. Otherwise the output is logic '0'.

The gate described above is also known as an **INCLUSIVE-OR** gate since it includes a '1' condition in the output column of the truth table when both A and B are '1' simultaneously (final row of table 3.3).*

As with other types of gate, the OR gate can have a fan-in greater than two; figure 3.5 illustrates a 5-input OR gate. The logical expression for the output from figure 3.5 is

$$F = A \text{ OR } B \text{ OR } C \text{ OR } D \text{ OR } E = A + B + C + D + E$$

The truth table for the above gate (which the reader may like to verify) has thirty-two rows; the output column of the truth table has only one logic '0', which occurs when all five variables have the value logic '0'.

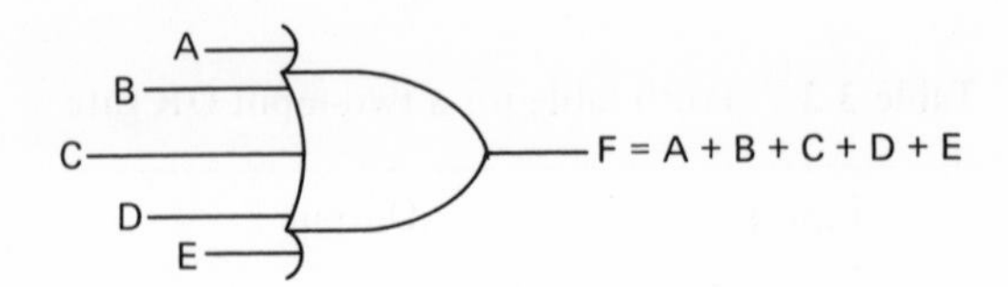

Figure 3.5

*Another type of gate, known as an **EXCLUSIVE-OR** gate, excludes the '1' in the output column under this set of input conditions. The first three rows of the output column of the truth table of the EXCLUSIVE-OR gate are identical to those of table 3.3, the fourth row of the output column containing a logic '0'.

3.6 The NOT Gate

A NOT gate gives a logic output signal which is NOT equal to the input signal. That is

$$\text{OUTPUT} = \text{NOT (INPUT)} = \overline{\text{INPUT}}$$

The bar written over the INPUT in the above expression represents the NOT function; the NOT operation is also described as **logic inversion** or **logical complementing.**

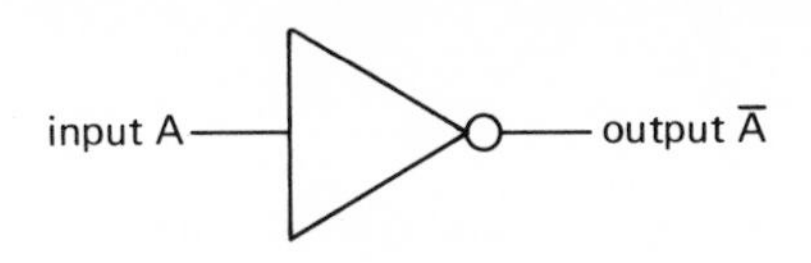

Figure 3.6

A symbol for a NOT gate is shown in figure 3.6; the small circle at the output represents the action of logical inversion. The reader will note that **the NOT gate has only one input line.** The truth table for the gate is given in table 3.4.

Table 3.4 Truth table for a NOT gate

Input A	Output NOT A = $\overline{A}$
0	1
1	0

3.7 The NAND Gate

A NAND gate can be regarded as an AND gate whose output is inverted by a NOT gate, as shown in figure 3.7a; the output from the gate is therefore the NOT function of the AND gate output. For a two-input NAND gate

$$\text{output, F} = \text{NOT (A AND B)} = \overline{\text{A.B}}$$

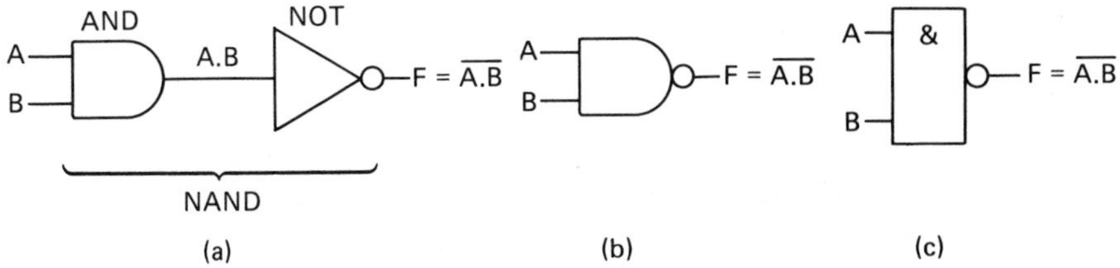

AND NOT A B A.B F = $\overline{\text{A.B}}$ NAND (a)

A B F = $\overline{\text{A.B}}$ (b)

A B & F = $\overline{\text{A.B}}$ (c)

Figure 3.7

Table 3.5 Truth table for a 2-input NAND gate

Inputs		Intermediate AND gate output (A.B)	Output
A	B		$F = \overline{A.B}$
0	0	0	1
0	1	0	1
1	0	0	1
1	1	1	0

The truth table for the two-input gate is given in table 3.5. The output from the AND section of the gate is given in the centre column of the truth table, each of these logical values being complemented by the NOT section of the gate. The truth table is summarised as follows

The output from a NAND gate is logic '1' whenever any input is logic '0'. Otherwise the output is logic '0'.

A NAND gate with a fan-in of five is illustrated in figure 3.8. The truth table for this gate has $2^5 = 32$ rows; the output column contains thirty-one 1's (which occurs when one or more of the inputs are logic '0') and one '0' (which occurs when all five inputs are logic '1' simultaneously).

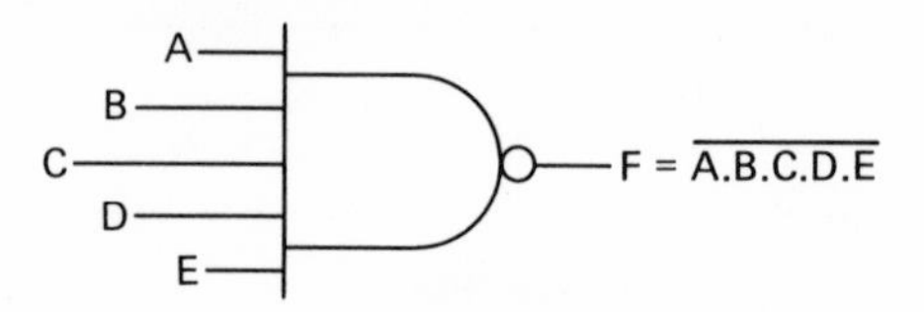

Figure 3.8

Readers are asked to note that a practical NAND gate is manufactured as a single circuit, and not as two separate gates (that is, the AND and NOT sections are not available as separate elements).

The NAND gate has several practical advantages over separate AND and NOT gates, and these are discussed in section 3.9.

3.8 The NOR Gate

The NOR gate can be regarded as consisting of an OR gate whose output signal is complemented by a NOT gate (see figure 3.9a). For a two-input NOR gate

$$\text{output, } F = \text{NOT (A OR B)} = \overline{A + B}$$

The truth table for the gate is given in table 3.6. The centre column of the table

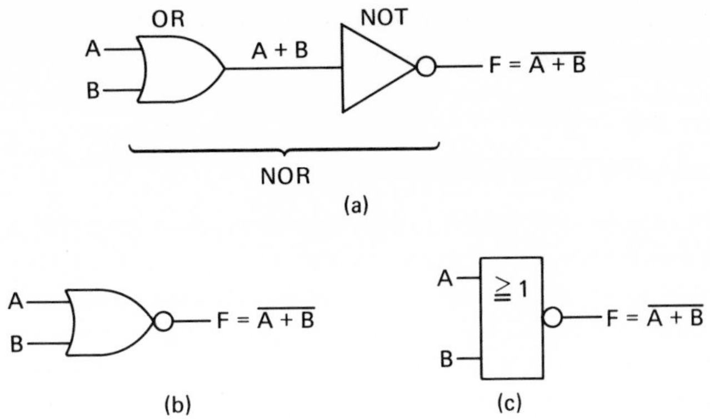

Figure 3.9

Table 3.6 Truth table for a 2-input NOR gate

Inputs		Intermediate OR gate output (A + B)	Output
A	B		$F = \overline{A + B}$
0	0	0	1
0	1	1	0
1	0	1	0
1	1	1	0

gives the output from the OR gate of the gate, which is inverted by the NOT gate. The truth table may be summarised as follows

The output from a NOR gate is logic '0' whenever any input is logic '1'. Otherwise the output is logic '1'.

A 5-input NOR gate is illustrated in figure 3.10; the output column of the truth table of this gate contains thirty-one 0's (which occur when one or more of the inputs are logic '1') and one logic '1' (which occurs when every input is logic '0' simultaneously).

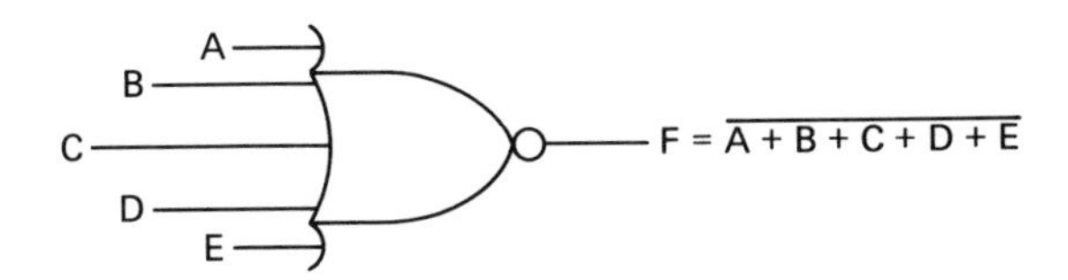

Figure 3.10

As with the NAND gate, the integrated circuit version of the NOR gate is a single circuit, and has significant advantages over the use of separate OR and NOT gates.

3.9 Universal Gates-NAND and NOR Gates

It is shown in the sections which follow, that all types of gate, that is, AND, OR, NOT, NOR and NAND can be replaced either by a group of NAND gates or by a group of NOR gates. An obvious advantage of this arrangement is that the user need only retain a stock of one type of gate (that is, either NAND gates or NOR gates) rather than stocks of several different types of gate. Furthermore, manufacturers can produce large quantities of one type of gate, with consequent reduction in production costs.

An apparent disadvantage of this arrangement is that several universal gates, say NAND gates, are needed to replace a single OR gate. However, using well-established circuit design methods*, it is possible to reduce or to **minimise** the number of NAND or NOR gates used to the same number of AND, OR and NOT gates needed to generate the same logic function.

The reader who is interested in logic circuit design principles should consult the further reading list at the end of this book.

3.10 Circuit Implementation Using NAND Gates

A circuit which develops the NOT function of a variable is shown in figure 3.11a.

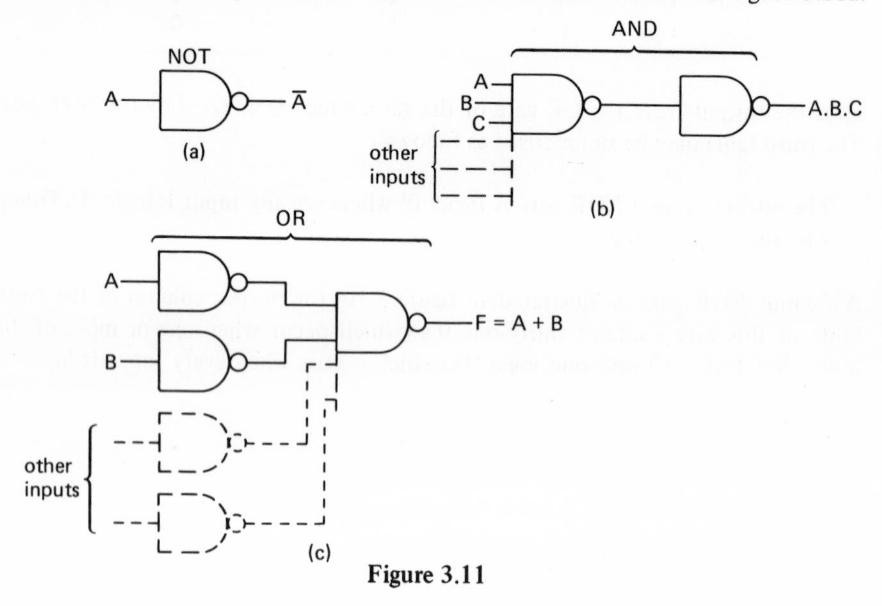

Figure 3.11

*See, for example, *Digital Electronic Circuits and Systems* by Noel M. Morris (Macmillan Press).

When a multiple-input NAND gate is used as a NOT gate, the question 'what must be done with the unused inputs' presents itself. Unused inputs are a potential source of trouble in logic systems, since induced voltages on these inputs can result in the logic circuits' malfunctioning. The reader is advised to connect each unused input line to a point in the circuit which does not affect the operation of the gate. In the case of a NAND gate, each unused input should be connected either to a logic '1' level or to a used input. The latter action has the disadvantage that it adds extra load to the driving gate.

The AND function of three inputs A, B and C is developed by the circuit in figure 3.11b. The AND function of more than three variables can be obtained if the first NAND gate in figure 3.11b has more than three input lines (see the dotted line connection).

The OR function of two variables A and B is generated by the circuit in figure 3.11c. If the OR function of more than two variables is needed, the circuit can be modified by adding the additional NAND gates shown by dotted lines.

In circuits (b) and (c), spare input lines are dealt with in the manner outlined in the first paragraph of this section.

A simple minimisation technique

When two NAND gates are cascaded in the manner shown in figure 3.12, both gates are redundant because the output signal has the same logical value as the input signal. That is, if the input is logic '1', then the output is logic '1'. Both gates may therefore be eliminated, and replaced by a piece of wire.

However, in microcomputer systems, a situation sometimes arises where it is necessary to deliberately introduce redundant gates using the circuit in figure 3.12. An example of this is outlined below.

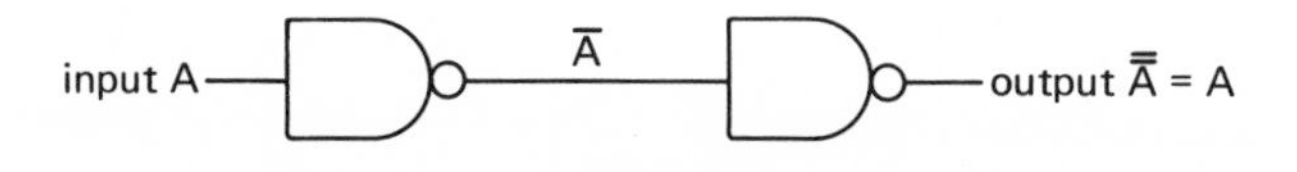

Figure 3.12

The number of logic gates that may be connected to the output of a 'driving' gate without causing it to malfunction is known as the **fan-out** of the gate. The fan-out capability of each output line of the majority of microprocessors is **one 'standard' TTL load.** If it is necessary to connect several logic gates to a microprocessor data bus line, then the data bus line must be protected or **buffered** from excessive load. One method of doing this is by the use of a simple circuit of the type in figure 3.12. The data bus would be connected to the input of figure 3.12, thereby ensuring that only one standard logic load is connected to that line, while the output from figure 3.12 can be used to 'drive' several standard TTL loads. The current handling capacity of a standard TTL gate is typically 16 mA, allowing the buffer circuit to drive a number of logic gates.

3.11 Circuit Implementation Using NOR Gates

A circuit which develops the NOT function of a variable is shown in figure 3.13a. In the case of NOR gates, unused input lines must either be connected to the logic '0' level or they may be connected to a used input line.

The OR function of three variables is developed by the circuit shown in full line in figure 3.13b. The circuit can be modified to generate the OR function of more than three variables by connecting the additional input signals to unused inputs of gate NOR 1, as shown by the dotted lines.

The AND function of two variables is generated by the circuit shown in full line in figure 3.13c. If the AND function of more than two variables is required, the additional input variables are connected to the input of NOR 2 as shown by the dotted lines.

Figure 3.13d illustrates a circuit in which both NOR gates are redundant, and both may be disconnected and replaced by a piece of wire. Alternatively, the circuit may be used as a buffer between a microprocessor and an external circuit.

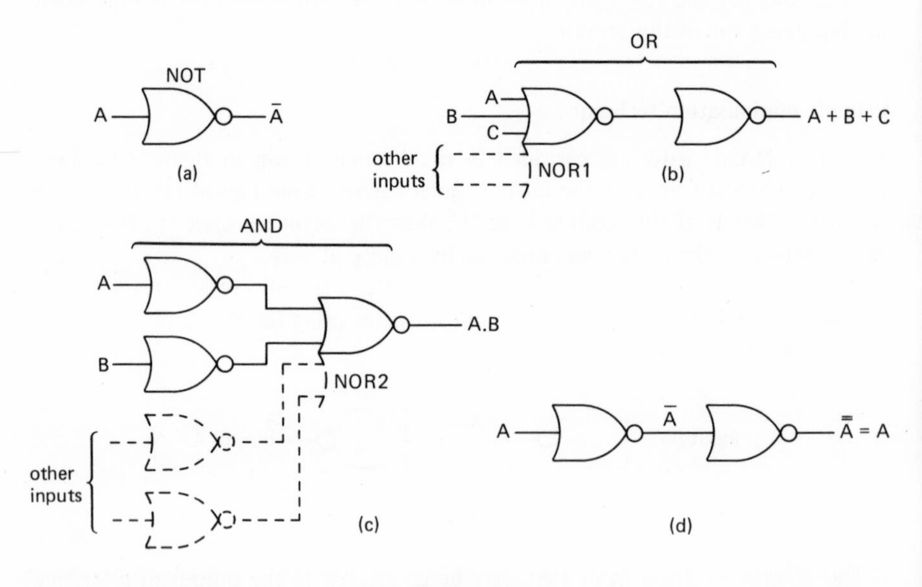

Figure 3.13

3.12 The EXCLUSIVE-OR Gate

The EXCLUSIVE-OR gate has the truth table in table 3.7. Note that output f is logic '1' either if (A = 0 AND B = 1) OR if (A = 1 AND B = 0). This relationship is expressed in logical algebraic form as

$$f = \overline{A}.B + A.\overline{B}$$

A circuit which generates the above logical expression using AND, OR and NOT

Table 3.7 Truth table for an EXCLUSIVE-OR gate

Inputs		Output
A	B	f
0	0	0
0	1	1
1	0	1
1	1	0

gates is shown in figure 3.14a (all-NAND and all-NOR versions are available in integrated circuit form).

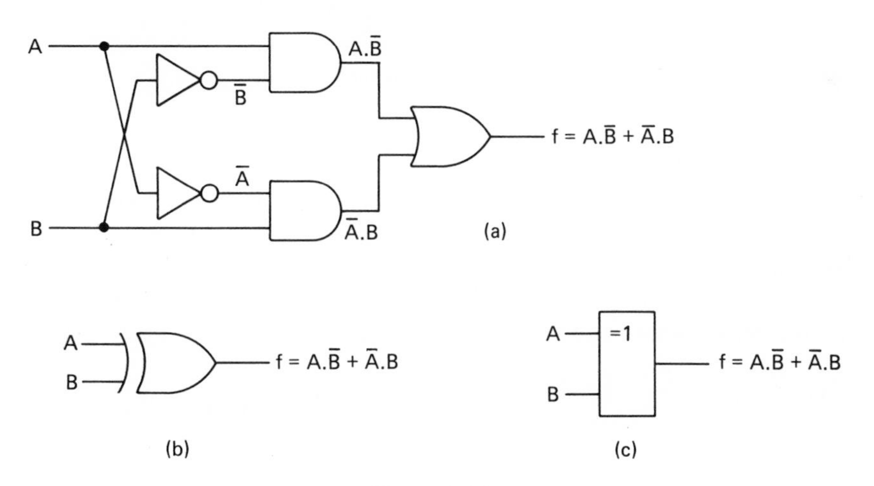

Figure 3.14

The EXCLUSIVE-OR gate is also known as a NON-EQUIVALENCE gate, since the output from the gate is logic '1' when signal A is NOT EQUIVALENT to signal B (that is, when A = 0 AND B = 1 OR when A = 1 AND B = 0). This gate is particularly useful when comparing two logic signals for equivalence, and is of great value in circuits which perform arithmetic operations such as addition and subtraction.

3.13 Flip-Flops, Latches and Flags

A **flip-flop** is a device used to store binary information. It has a number of **input signal lines** (usually one or two, the latter being more common) along which the data to be stored is transmitted. There are also a number of control lines; the signal on one of these lines (the **clock signal**) controls the movement of data through the flip-flop. Other control signal lines can be used to cause the output of the flip-flop either to be forced to the logic '0' state (this is known as the **reset signal** or as the

clear signal or as the **preclear signal**) or to be forced to logic '1' (this is known as the **set signal** or **preset signal**).

Most flip-flops have two **output lines**; one is known as the Q-output and the other as the NOT-Q or $\overline{Q}$ output. Under normal operating conditions, the signals on the Q and $\overline{Q}$ lines are complementary; that is, if Q = 1 then $\overline{Q}$ = 0 and vice versa.

A flip-flop is used to 'latch' a binary level, and for this reason they are sometimes described as **latches**. Also, in a microprocessor, a **flag** is a flip-flop which can either be 'set' to logic '1' or can be 'reset' to logic '0' in response to an operation in the microprocessor. One such flag is the **zero flag**, which is set to logic '1' when the value stored in the accumulator is zero; otherwise the flag is reset to logic '0'.

The description given below is necessarily brief, detailed information being given in more specialised textbooks (see, for examples, the reading list at the end of the book).

S–R flip-flop

The simplest form of flip-flop is the S–R (set–reset) flip-flop in figure 3.15a; its implementation in NOR gate form is given in figure 3.15b. The operation of the S–R flip-flop is described by table 3.8, in which Q_n is the state of output Q after n operations (where n is an integer). Q_{n+1} is the state of output Q after the $(n + 1)$th set of input conditions have been applied.

To illustrate the effect of the first row of the truth table, consider the case when the nth set of input conditions has resulted in Q being set to logic '1' (that is,

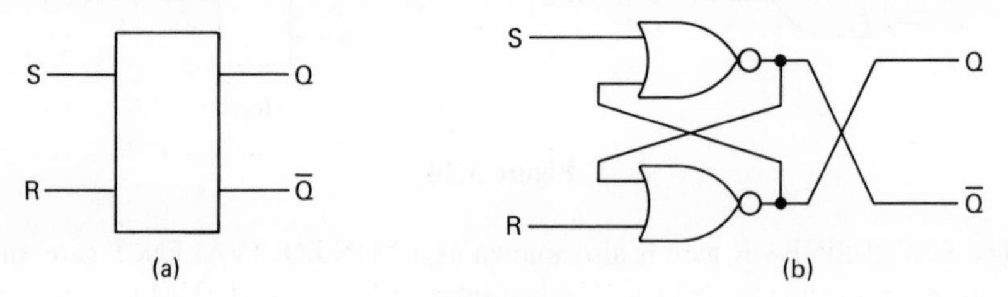

Figure 3.15

Table 3.8 Truth table for an S–R flip-flop

Inputs		Output	Comment
S_{n+1}	R_{n+1}	Q_{n+1}	
0	0	Q_n	'Store' or 'memory' state
0	1	0	'Reset'
1	0	1	'Set'
1	1	X	'Don't know'

$Q_n = 1$). When both S and R are changed to logic '0' that is, $S_{n+1} = 0$ and $R_{n+1} = 0$, output Q remains unchanged at logic '1'. That is $Q_{n+1} = Q_n$.

If the input conditions are changed so that S = 0, R = 1 (second row of the truth table), output Q is reset to '0'; in this case $Q_{n+1} = 0$. When the input conditions are changed to S = 1, R = 0 (third row of the truth table), output Q is set to '1'; in this case $Q_{n+1} = 1$.

The output condition in the final row of table 3.8 is given as X, where X is described as a 'don't know' condition. Strictly speaking, this is not quite correct since, using the NOR gate implementation of the S–R flip-flop, it will be seen that when S = 1 and R = 1 simultaneously, then Q = 0 and $\overline{Q} = 0$ simultaneously (remember, when a logic '1' is applied to any input of a NOR gate, its output is logic '0'). However, with both S and R inputs initially at logic '1', if *both* inputs are simultaneously changed from logic '1' to '0', there is a 'race' between the two gates to reach an output of logic '1'; whichever gate reaches the logic '1' level first forces (via the feedback link between the gates) the output of the other gate to be logic '0'. In this case we can truly say that we don't know which output will be '1' and which '0' until the race has finished.

Level triggered master–slave J–K flip-flop

The master–slave J–K flip-flop contains two cascaded stages known respectively as the master stage and the slave stage. Although the J–K flip-flop is more complex than the S–R flip-flop, it has the advantage that the 'race' condition which occurs when both the S- and R-input lines simultaneously change from '1' to '0' is eliminated. Moreover, it has a truth table which is generally similar to that of the S–R flip-flop; for this reason, the S–R flip-flop can be replaced by the J–K element.

This flip-flop has two input lines, namely the J-line and the K-line, and two output lines Q and $\overline{Q}$ (see figures 3.16a). In addition it has a clock signal line, and may have 'set' and 'clear' lines which allow the user either to 'set' or 'clear' the Q-output *when the clock signal is at logic '0'.* In many instances the set and clear lines require an 'active low' signal, that is a logic '0' implements the set and clear actions. In the latter case, these control lines are described as $\overline{\text{set}}$ and $\overline{\text{clear}}$ lines; the reader is advised to consult specification sheets in respect of individual IC packages.

Movement of data through the J–K flip-flop is controlled by the clock signal (figure 3.16b), which has a rectangular voltage–time waveform. When the clock signal changes from logic '0' to logic '1', the data at the J- and K-input lines is connected to the master stage of the flip-flop; the data in the slave (output) stage is unchanged at this time. When the clock signal changes from logic '1' to logic '0', the data stored in the master stage is transmitted from the master stage to the slave stage, at which time the data appears at the output terminals Q and $\overline{Q}$.

It therefore takes one clock pulse to cause signals applied to the J- and K-input lines to be transmitted to the output terminals.

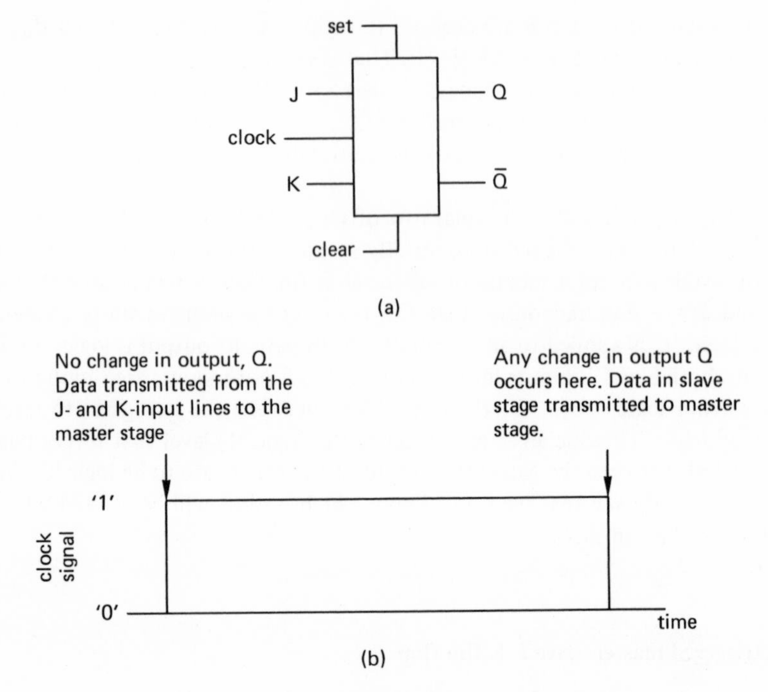

Figure 3.16

This type of element is known as a **level-triggered flip-flop** since data is transmitted through the flip-flop when the voltage level of the clock signal changes.

The truth table of the J-K flip-flop is given in table 3.9. Inputs J_{n+1} and K_{n+1} are the $(n + 1)$th set of input signals applied *prior* to the $(n + 1)$th clock pulse; Q_{n+1} is the state of output Q after the completion of the $(n + 1)$th clock pulse. The reader will note from table 3.9 that the first three rows of the J-K truth table replace the first three rows of the S-R truth table (table 3.8) if we regard the J-input as being equivalent to the S-input of the S-R element, and the K-input as equivalent to the R-input.

Table 3.9 Truth table of a J-K flip-flop

Inputs		Output	Comment
J_{n+1}	K_{n+1}	Q_{n+1}	
0	0	Q_n	'memory' state
0	1	0	'reset' state
1	0	1	'set' state
1	1	$\overline{Q}_n$	'toggle' state

In the case where $J_{n+1} = 1$ and $K_{n+1} = 1$ simultaneously (see the final row of table 3.9), output Q changes state *after* the completion of each clock pulse. For instance, if Q = 0 initially, then if J = K = 1, a sequence of clock pulses cause output Q to change sequentially as follows: 0, 1, 0, 1, 0, 1, ... This mode of operation is one which is widely used in binary counting circuits.

Edge-triggered D flip-flop

The D flip-flop (figure 3.17) has a single input line (the D-line), and two output lines (Q and $\bar{Q}$). The operation of the flip-flop is controlled by a clock signal; in addition the flip-flop may have 'set' and 'clear' control lines. The D flip-flop can perform the functions of the J-K flip-flop, its advantage over the J-K element being that it only has one input line.

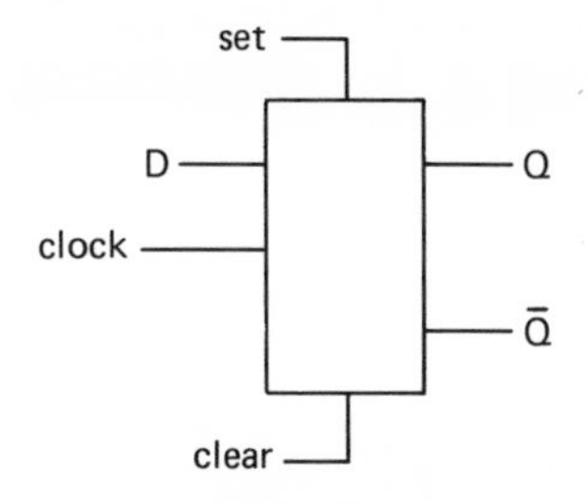

Figure 3.17

Commercially available D flip-flops are **edge-triggered** types, in which the data applied to the D-line is transmitted to the output terminal within a few nanoseconds of the clock pulse changing from logic '0' to logic '1'. That is, the data transfer can be regarded as occurring on the application of the **leading edge** of the clock pulse.

The truth table of the flip-flop is given in table 3.10. In the table, D_{n+1} is the logic signal applied to the D-line *before* the $(n + 1)$th clock pulse is applied to the flip-flop. Q_{n+1} is the state of output Q *after* the leading edge of the $(n + 1)$th pulse has been applied.

Table 3.10 Truth table of the D flip-flop

Input D_{n+1}	Output Q_{n+1}
0	0
1	1

The D flip-flop is often used as a **data latch** element as follows. The data which is output from a microprocessor may only be available on the data bus for a fraction of a microsecond. It is necessary to 'latch' this data by means of, say, a D flip-flop, if the data is to be used by an external display device such as a lamp or light-emitting diode.

3.14 Integrated Circuits

Integrated circuits (ICs) used in association with microprocessors are manufactured in what is known as **planar monolithic** form; that is to say, they are constructed in a very thin slice of silicon (a 'chip') in a flat or plane form. The transistor circuit in figure 3.18a is manufactured in the monolithic form in figure 3.18b.

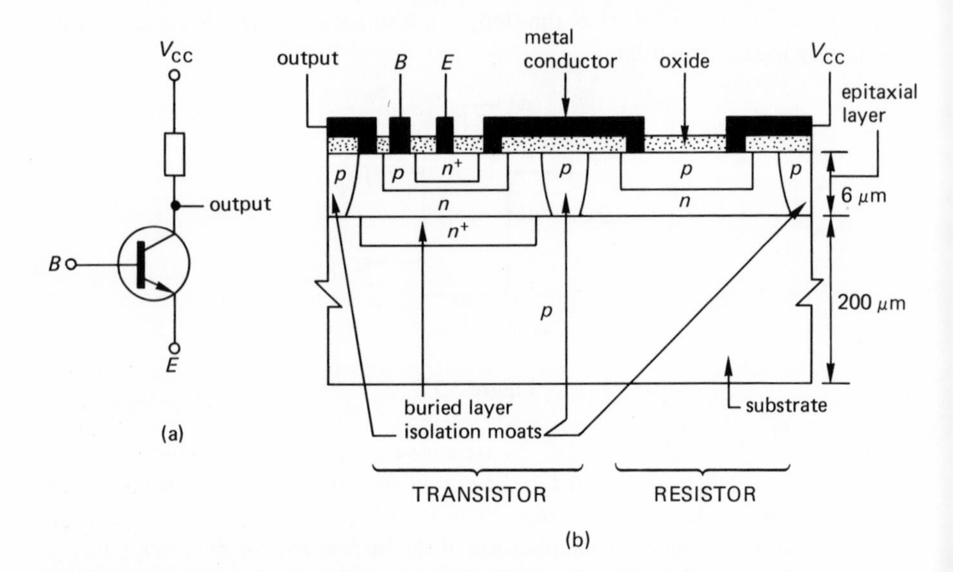

Figure 3.18 Monolithic integrated circuit

After manufacturing, the silicon chip is packaged in one of three basic forms:

(1) 'top hat' type of canister (or 'cans')
(2) flatpacks
(3) dual-in-line (DIL) packs

Outline diagrams of the three types are shown in figure 3.19. Types 1, 2 and 3 above correspond to diagrams (a), (b) and (c) in the figure. The most popular form is the DIL, which is widely used in industrial and commercial equipment; microprocessors are manufactured in a 40-pin DIL form.

The gates, flip-flops and other devices associated with microprocessors are fabricated as **large-scale integrated** (LSI) circuits, and are encapsulated in DIL form.

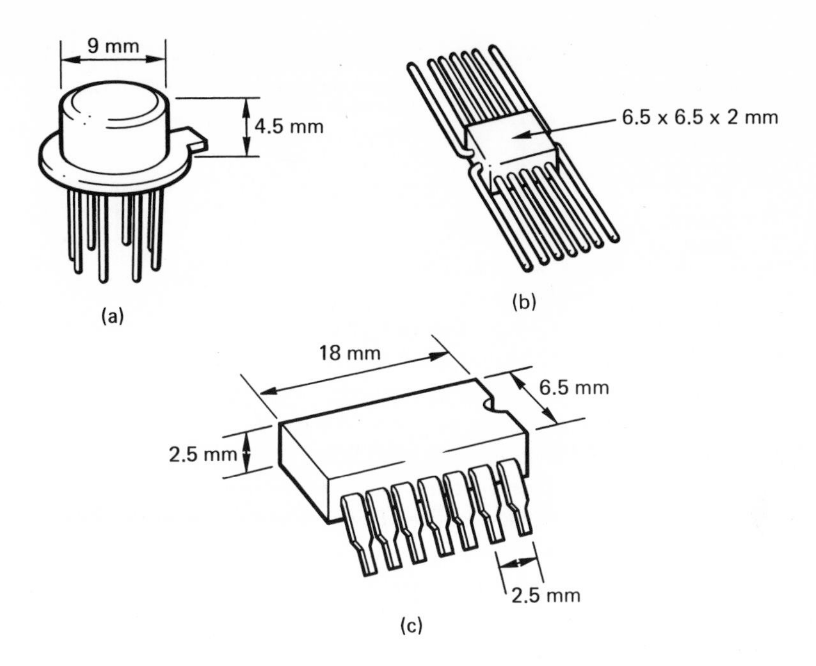

Figure 3.19 Methods of packaging integrated circuits

The logic families used are typically **transistor-transistor logic (TTL)**, or **metal-oxide-silicon (MOS) field-effect transistor logic** or **integrated injection logic (I^2L)** gates. For more information on these logic families see the further reading list given at the end of the book.

As mentioned earlier, it is usually the case that the output current available from a microprocessor bus line can 'drive' only one *standard* TTL gate input. However, it is often necessary to connect the inputs of several gates to a single data bus line, so that the fan-out of the bus line must be increased. This requirement can be satisfied either (a) by using a buffer circuit (see figures 3.12 and 3.13d) or (b) by using gates whose input terminal current is less than that of a standard TTL gate. MOS and I^2L gates fall into the latter category, and several of these gates can be connected directly to a microprocessor bus line without the use of a buffer circuit. Additionally, a sub-family of TTL, known as **low-power TTL**, also requires a low input current. Several of these gates may be connected directly to a microprocessor bus line without buffering.

3.15 Three-State Logic Gates

A **three-state gate** or **tri-state gate** is one in which the output line may take one of three conditions, namely (a) logic '0', (b) logic '1', (c) open-circuit. The basis of a three-state NAND gate is shown in figure 3.20a. The reader will note that a control line, known as an **enable** line, is used to control the state of switch S.

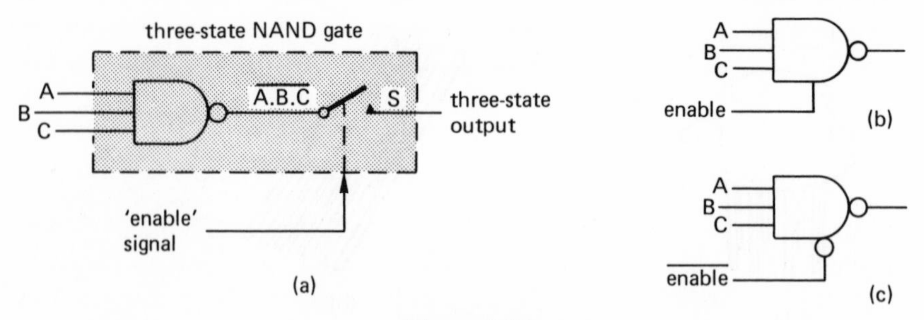

Figure 3.20

When the enable line is activated by a suitable logic signal (see below), switch S closes and the logic signal generated by the NAND gate (either a '0' or a '1') is transmitted to the output of the three-state gate. When the 'enabling' signal is removed, the gate is 'disabled'; this causes switch S to be opened, and the output line is open-circuited, that is, the output line is isolated from the logic element.

Depending on the gate design, the enabling signal may either be high (logic '1') or low (logic '0'). The circuit symbol for a 'high' enable gate is shown in figure 3.20b, and for a 'low' enable gate in figure 3.20c.

Three-state operation is not limited to individual gates, and is widely used in many other logic elements associated with microprocessors. Included in these are latches (flip-flops), buffer gates, inverting gates, input/output devices, etc.

It is common practice in microprocessor equipment to transmit digital data from several different sources along a single wire; that is, the wire is shared between the signal sources. This process is known as **multiplexing**; three-state logic devices are very valuable in these applications. A simple multiplexing system is shown in figure 3.21. Signals A and B are to be transmitted sequentially along the common link or **bus** from the left-hand side of the diagram to the right-hand side. When signal A is to be transmitted, gates A1 and A2 are enabled by a logic '1' applied to control line E_A. Gates B1 and B2 are disabled at this time by applying a logic '0' signal to control line E_B. When signal B is to be transmitted, the signal on E_A is forced to logic '0', while that on E_B becomes logic '1'.

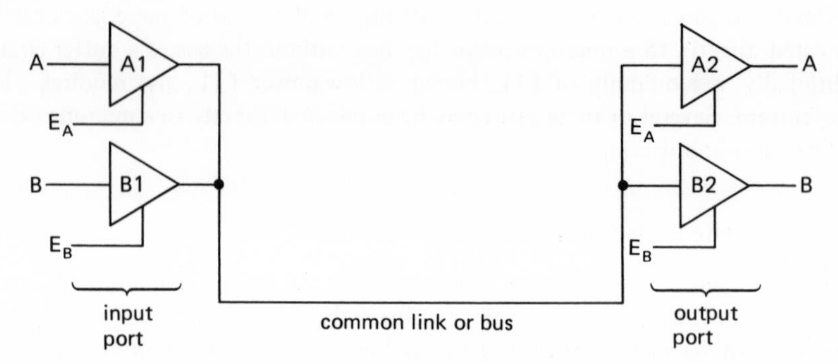

Figure 3.21

As mentioned in section 3.14, microprocessors are generally capable of driving only a single 'standard' logic load. While this may be adequate for a small system, it is often necessary to buffer the microprocessor (and also the memory elements) when extra load is connected. Moreover, in the case of the data bus, it may be necessary to provide a bidirectional drive; that is, the data bus not only accepts data from an input device via the three-state buffer, but the buffer must also be capable of transmitting data to the 'outside' world. The logic block diagram of a 4-bit data **bidirectional bus driver** is shown in figure 3.22. The circuit uses eight three-state buffers; the buffers labelled A allow transfer of logic signals from the data bus to an output device, and the buffers labelled B permit transfer of logic signals from an 'input' device to the data bus. Although the buffer gates A and B in figure 3.22 are non-inverting, inverting versions are also manufactured.

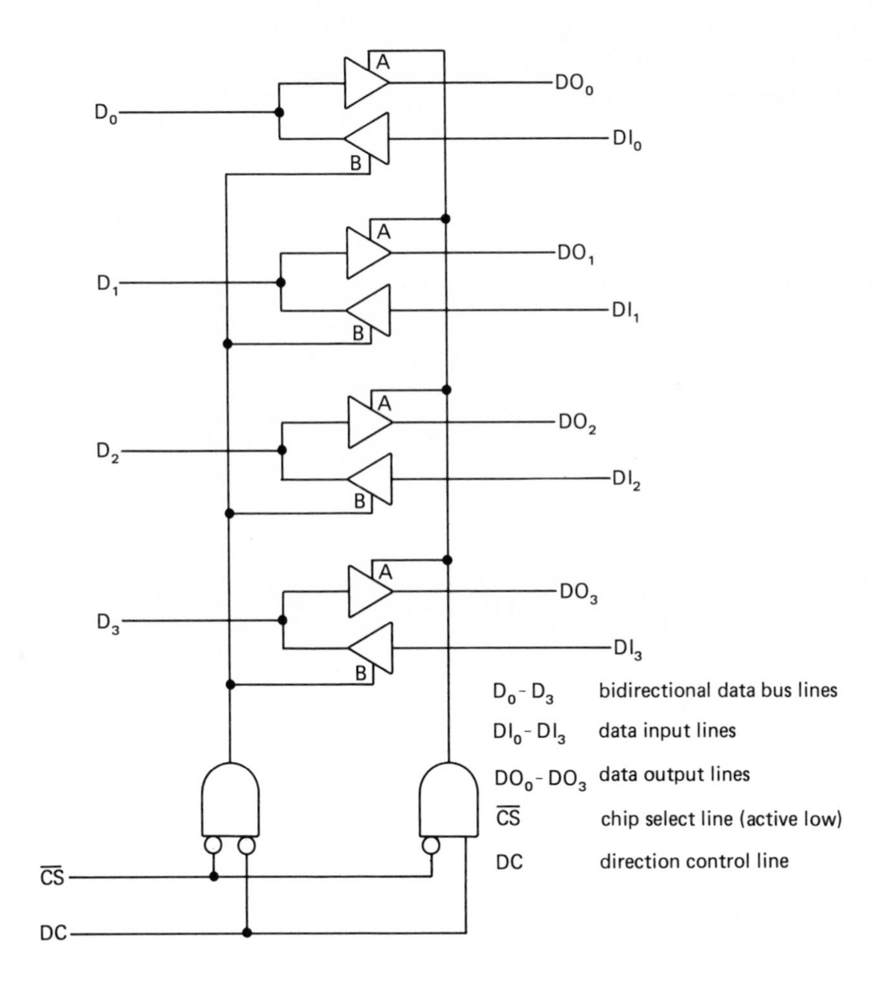

Figure 3.22

In order to 'select' the chip (either group A or group B gates), the $\overline{CS}$ line must be held low. When $\overline{CS} = 1$, all eight gates are driven into their high output impedance state. To select group A gates (these connect the D lines to the DO lines, that is, D_0 to DO_0, etc), it is necessary for $\overline{CS}$ to be logic '0' AND for DC (direction control) to be logic '1'. To select group B gates (which connect the DI lines to the D lines), it is necessary for $\overline{CS}$ to be '0' AND for DC to be logic '0'.

3.16 Decoders

The 'chip-select' signals required by many microprocessor support chips are, very frequently 'active low' signals; that is, the chip-select input is labelled as a $\overline{CS}$ or a $\overline{DS}$ line. To select an individual support chip in a microprocessor system, it is often necessary to decode from 'active high' signals on the address bus to an 'active low' which selects the appropriate support chip.

A simple analogy may be drawn between an electronic decoder and a set of railway points. The set of points directs the train on the incoming line to the appropriate track, much as the decoder directs or addresses the incoming data to the appropriate output.

A simplified block diagram of a **1-out-of-4 decoder** (also known as a **2-to-4 line decoder**) is shown in figure 3.23. The signals on the address bus lines A_0 and A_1 select the appropriate output line. The truth table for the chip in figure 3.23 is given in table 3.11.

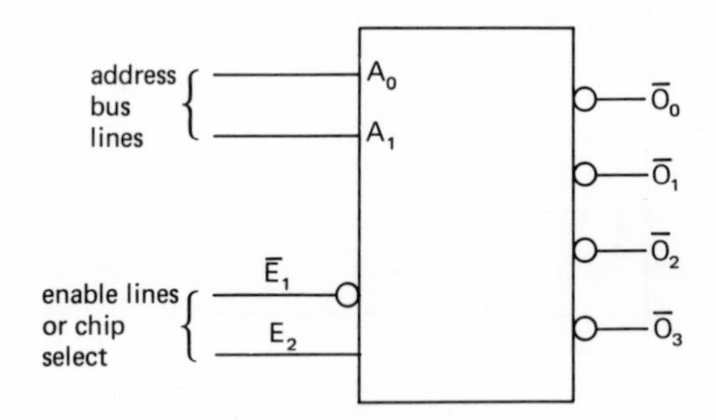

Figure 3.23

The chip in figure 3.23 is selected when $\overline{E}_1 = 0$ AND $E_2 = 1$ simultaneously, the AND function being generated inside the chip. It is often the case that more than one signal is required to ensure chip selection. The use of an active low ($\overline{E}_1$) and an active high (E_2) control line gives additional flexibility, since the device(s) providing the control signals may provide either logic level. Although two chip-select signals are apparently needed to correctly select the chip in figure 3.23, the requirement can be reduced to one chip-select signal as follows. If the output signal from the device providing the chip-select signal is active 'low', it is connected to the $\overline{E}_1$ line;

Table 3.11 Truth table for a 1-out-of-4 decoder

Inputs				Outputs				
A_0	A_1	$\overline{E}_1$	E_2	O_0	O_1	O_2	O_3	
0	0	0	1	0	1	1	1	
0	1	0	1	1	0	1	1	
1	0	0	1	1	1	0	1	
1	1	0	1	1	1	1	0	
X	X	X	0	1	1	1	1	X = '0' or '1'
X	X	1	X	1	1	1	1	

the E_2 line can be permanently connected to a logic '1' signal such as $+V_{CC}$. If the chip-select signal is active 'high', it is connected to line E_2; the $\overline{E}_1$ line can be permanently connected to a logic '0' signal such as zero volts.

As shown in the truth table, when $\overline{E}_1 = 0$ AND $E_2 = 1$, the output signal on the selected line (given by the decimal weight of the binary value on address lines A_1 and A_2) is active 'low'. Thus when $A_0 = 0$ AND $A_1 = 0$, output line zero ($\overline{O}_0$) is selected. When $A_0 = 1$ AND $A_1 = 0$, output line 'one' ($\overline{O}_1$) is selected, etc; the logic level on the selected line falls to logic '0', the other outputs remaining at logic '1'. When $\overline{E}_1 = 1$ or when $E_2 = 0$, all outputs are 'high', that is, they are deselected, whatever signals are applied to A_0 and A_1.

The outputs $\overline{O}_0$ - $\overline{O}_4$ are connected to the $\overline{\text{chip select}}$ of four chips in the microcomputer system (not shown).

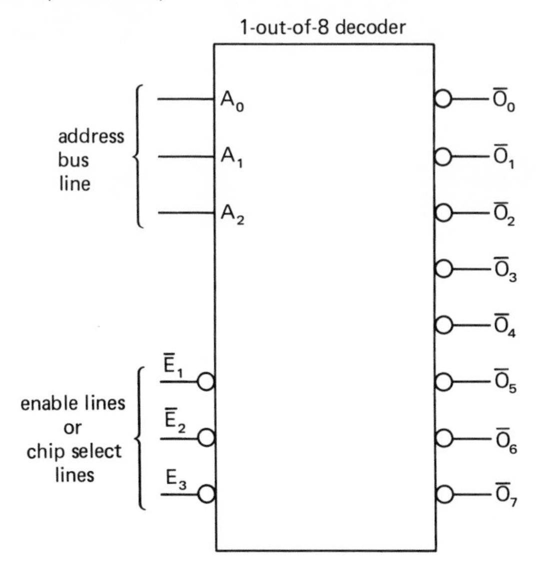

Figure 3.24

Up to eight chips can be selected using a **1-out-of-8 decoder** of the type in figure 3.24, its truth table being given in table 3.12. Typical of this type is the 8205 IC, which uses Schottky bipolar technology. The eight output lines ($\overline{O}_0$ - $\overline{O}_7$) are addressed by three address bus lines (A_0, A_1 and A_2), and the chip has three chip-select lines $\overline{E}_1$, $\overline{E}_2$ and E_3 (either one, two or all three of these lines may be used, the unused line(s) being connected in the way described for the 1-out-of-4 decoder). This decoder permits any one of eight support chips to be individually selected.

Table 3.12 Truth table for a 1-out-of-8 decoder

Address			Enable			Outputs							
A_2	A_1	A_0	E_3	$\overline{E}_2$	$\overline{E}_1$	0	1	2	3	4	5	6	7
0	0	0	1	0	0	0	1	1	1	1	1	1	1
0	0	1	1	0	0	1	0	1	1	1	1	1	1
0	1	0	1	0	0	1	1	0	1	1	1	1	1
0	1	1	1	0	0	1	1	1	0	1	1	1	1
1	0	0	1	0	0	1	1	1	1	0	1	1	1
1	0	1	1	0	0	1	1	1	1	1	0	1	1
1	1	0	1	0	0	1	1	1	1	1	1	0	1
1	1	1	1	0	0	1	1	1	1	1	1	1	0
X	X	X	0	X	X	1	1	1	1	1	1	1	1
X	X	X	X	1	X	1	1	1	1	1	1	1	1
X	X	X	X	X	1	1	1	1	1	1	1	1	1

X = '0' or '1'

A variety of other decoders exist, including the 74LS154 **1-out-of-16 decoder** (or 4-to-16 line decoder). This chip has four address lines which select any one of sixteen output lines; it also has two active low enable control lines.

3.17 A Practical I/O Port

A simplified version of a practical input/output port (based on the Intel 8212 chip) is shown in figure 3.25. This port satisfies the requirements of the system described in chapter 1. It deals with up to eight sets of data input signals, which are connected to the input pins DI_0 - DI_7, the output from the port being taken from the data output pins DO_0 - DO_7. The function of the pins is as follows

DI_0 - DI_7	data input
DO_0 - DO_7	data output
$\overline{DS1}$, DS2	device select
MD	definition of operating mode
STB	strobe
$\overline{CLR}$	clear (active low)

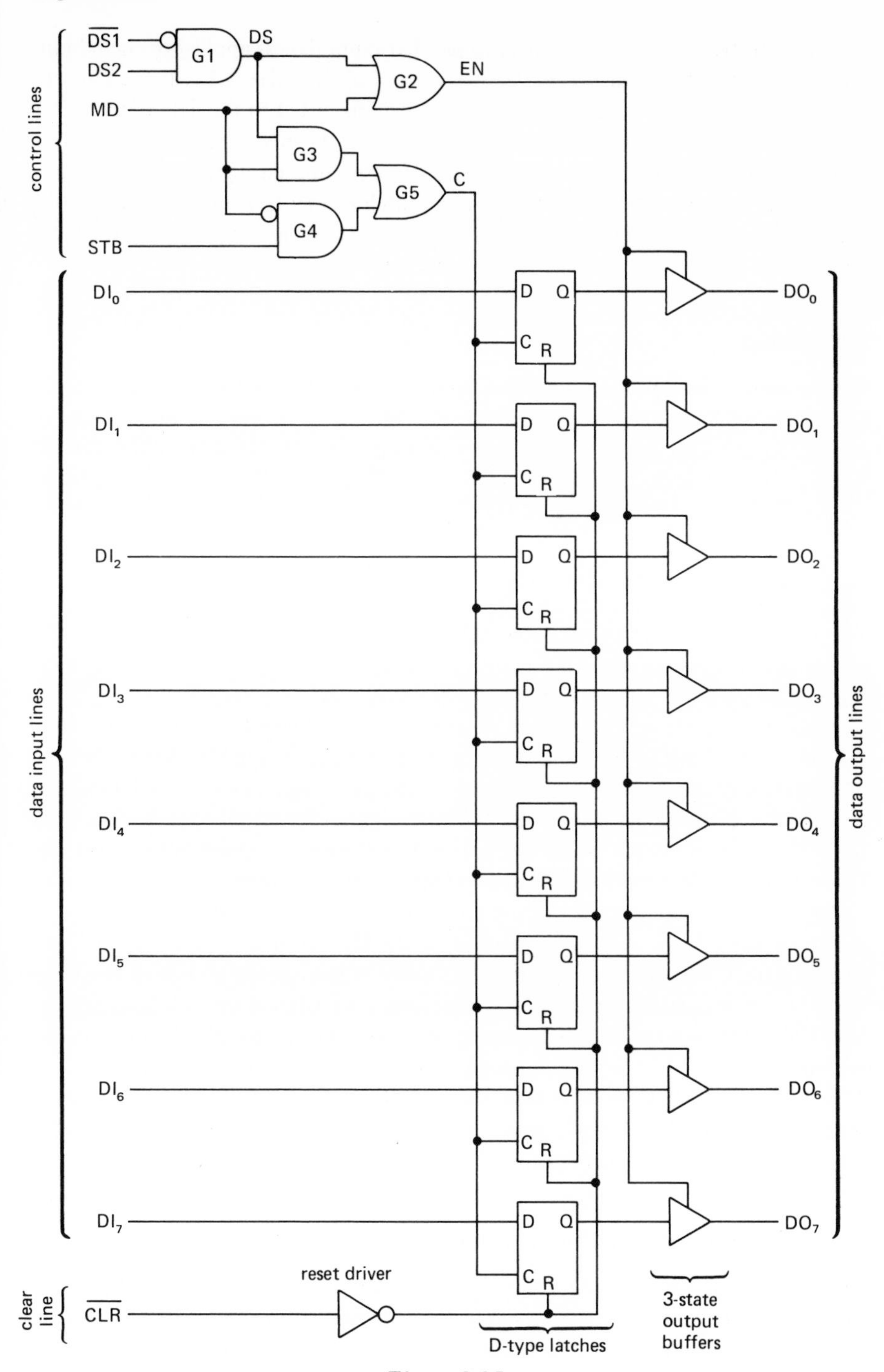
control lines
$\overline{DS1}$
DS2
MD
STB
G1
DS
G2
EN
G3
G4
G5
C
data input lines
DI_0
DI_1
DI_2
DI_3
DI_4
DI_5
DI_6
DI_7
D Q
C R
DO_0
DO_1
DO_2
DO_3
DO_4
DO_5
DO_6
DO_7
data output lines
clear line
$\overline{CLR}$
reset driver
D-type latches
3-state output buffers

Figure 3.25

Referring to figure 3.25, the reader will see that input data is applied to the DI pins, this data being latched in D-type latches when a clock pulse (C) is applied by gate G5 to the clock line of the latches. The data is transmitted to the data output (DO) pins when the three-state buffers are enabled when the EN signal at the output of G2 is logic '1'. When EN = 0, the output buffers are in their high impedance (open-circuit) output state. The Q-output of every latch is reset to zero when a logic '0' is applied to the $\overline{\text{CLR}}$, line; if it is not necessary to clear the latches at any stage, the $\overline{\text{CLR}}$ line is connected to a logic '1' level (such as $+V_{CC}$).

Device select

Two separate 'device-select' lines are available; in order to select the I/O chip, a logic '0' must be applied to the $\overline{\text{DS1}}$ input AND a logic '1' must simultaneously be applied to the DS2 input. Thus the device select signal (DS) at the output of G1 is given by the logic combination $\text{DS} = \overline{\text{DS1}}.\text{DS2}$; when DS = 1, the three-state output buffers are enabled via gate G2.

Mode signal

The signal on this line is used not only to control the state of the output-buffer-enable signal (EN) but also to establish the source of the clock-input signal (C) to the data latches.

When MD = 1 the output buffers are enabled (EN = 1) irrespective of the state of the device-select signal (DS). At the same time, this signal enables gate G3 and inhibits gate G4; consequently, the source of the clock signal (C) to the data latches is from the device-select signal DS. Thus data is transferred into the latches when the I/O chip is selected. When the port is to be used as a **gated output port** (see figure 3.26), the MD line is connected to logic '1', that is, to $+V_{CC}$.

When MD = 0 the output buffers are enabled only when the device is selected, that is, when DS = 1. A logic '0' applied to the MD terminal inhibits gate G3 and enables gate G4; hence the strobe (STB) signal is used as the source of the clock signal (C) to the data latches. In this case, data is transferred into the latches when STB = 1. When used as a **gated input port** (see figure 3.26), the MD line is connected to logic '0', that is, to ground, and the STB line is connected to logic '1'.

Note: The 8212 I/O port provides an additional output signal known as the **interrupt request signal** ($\overline{\text{INT}}$, active low); this signal is derived from the control signals. The function of this signal is described in chapter 9.

3.18 Basic I/O Connections to a Microprocessor Bus System

The connections between the basic I/O port described in section 3.17 and the CPU bus system are shown in figure 3.26. In the case of the **input port**, the MD is held low and the STB line is held high. As a result, the clock signal (C) in the input port (see figure 3.25) is high. Consequently, the data input signals (DI) are transferred

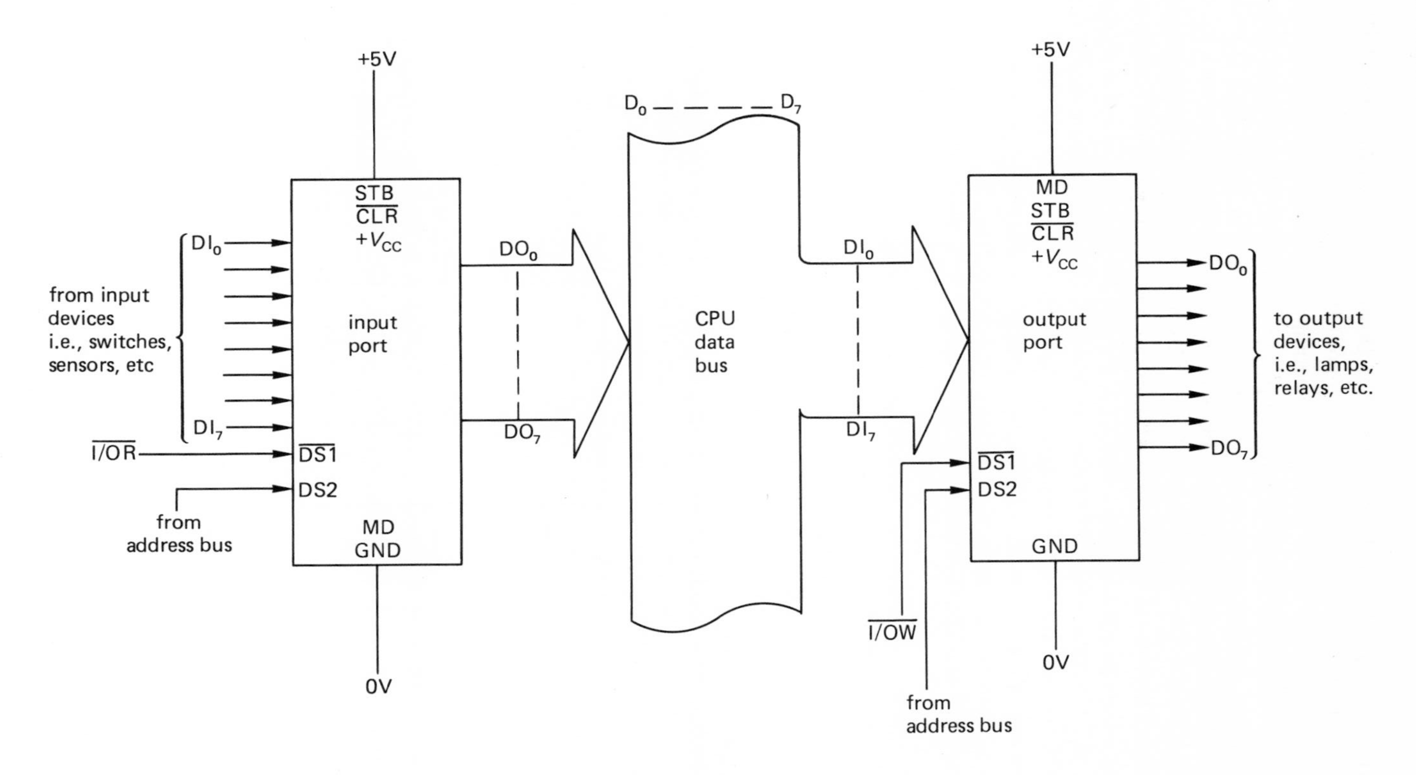
from input devices i.e., switches, sensors, etc
DI_0
DI_7
$\overline{I/OR}$
from address bus
+5V
$\overline{STB}$
$\overline{CLR}$
$+V_{CC}$
input port
$\overline{DS1}$
DS2
MD
GND
0V
DO_0
DO_7
D_0
D_7
CPU data bus
DI_0
DI_7
$\overline{I/OW}$
from address bus
+5V
MD
STB
$\overline{CLR}$
$+V_{CC}$
output port
$\overline{DS1}$
DS2
GND
0V
DO_0
DO_7
to output devices, i.e., lamps, relays, etc.

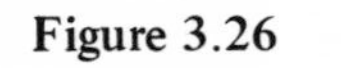
Figure 3.26

directly to the Q-outputs of the data latches. However, the data in the latches is not transmitted to the data output (DO) lines until the output buffers of the chip are enabled. This occurs when the device selection is 'true', that is, when $\overline{DS1}$ is low AND DS2 is high; for this condition to apply, the input/output read signal ($\overline{I/OR}$) is low AND the signal from the address bus which is applied to DS2 is high. The timing of the waveforms produced by the CPU is discussed in chapter 4.

In the case of the **output port** in figure 3.26, both the MD and the STB lines are held high. This ensures that the output buffers are permanently enabled; consequently, the Q-outputs of the latches are directly applied to the DO lines of the output port. When the CPU executes an I/O write instruction, the $\overline{I/OW}$ control line is forced low AND, at the same time, the signal applied to the DS2 line of the chip from the address bus is high. This causes the clock signal (C) inside the output port (see also figure 3.25) to become logic '1', resulting in the data on the CPU data bus (D_0 - D_7) being latched into the flip-flops in the output port.

3.19 Priority Encoders

Yet another type of MSI chip used in microprocessor systems is the priority encoder. This chip enables the user to rank a number of input signals in order of importance, the binary output signal from the priority encoder indicating the 'most important' signal applied at any given time.

A diagram of a popular form of 8-bit priority encoder, the 74148 chip, is shown in figure 3.27; its truth table is given in table 3.13. For the chip to function normally,

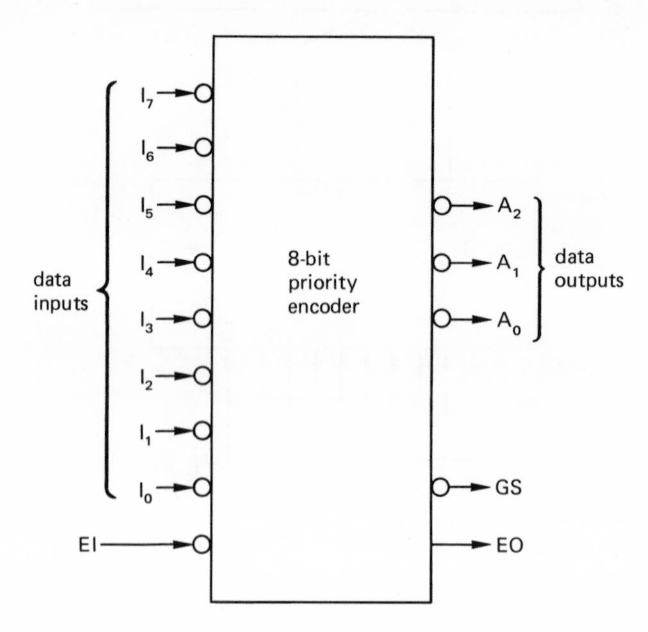

Figure 3.27

Table 3.13 Truth table for an 8-bit priority encoder

Inputs									Outputs				
EI	I_0	I_1	I_2	I_3	I_4	I_5	I_6	I_7	A_2	A_1	A_0	GS	EO
0	X	X	X	X	X	X	X	0	0	0	0	0	1
0	X	X	X	X	X	X	0	1	0	0	1	0	1
0	X	X	X	X	X	0	1	1	0	1	0	0	1
0	X	X	X	X	0	1	1	1	0	1	1	0	1
0	X	X	X	0	1	1	1	1	1	0	0	0	1
0	X	X	0	1	1	1	1	1	1	0	1	0	1
0	X	0	1	1	1	1	1	1	1	1	0	0	1
0	0	1	1	1	1	1	1	1	1	1	1	0	1
0	1	1	1	1	1	1	1	1	1	1	1	1	0
1	X	X	X	X	X	X	X	X	1	1	1	1	1

X = '0' or '1'

the Enable Input (EI) line must be low. The eight input signals (active low) are connected to the I_0 - I_7 lines. The output signals appear on the A_2 (m.s.b.), A_1 and A_0 (l.s.b.) lines. Input I_0 has the lowest priority while input I_7 has the highest priority. Thus a logic '0' on input I_4 overrides a logic '0' on input I_3.

When any input line is selected, the EO (Enable Output) line is forced to logic '1', EO is low only when none of the inputs is selected AND EI is low. The GS (Gated Strobe) line is forced low when one input (or more) is selected AND EI is low.

The IC package described above can be used in priority interrupt circuits (see chapter 9, section 9.8) and also as a keyboard encoder for microcomputer systems.

PROBLEMS

3.1 State the name of each gate (G1–G4) in table 3.14.

Table 3.14

Inputs		Output			
A	B	G_1	G_2	G_3	G_4
0	1	0	1	0	1
0	0	0	0	1	1
1	0	0	1	0	1
1	1	1	1	0	0

3.2 Design a logic network using only NAND gates which checks for illegal code groups in the 8421 BCD code in table 2.2.

3.3 Design a logic network which satisfies table 3.15 using (a) only NOR gates, (b) only NAND gates.

Table 3.15

Inputs			Output	Inputs			Output
A	B	C	f	A	B	C	f
0	0	0	0	1	0	0	1
0	0	1	1	1	0	1	0
0	1	0	1	1	1	0	1
0	1	1	1	1	1	1	1

3.4 Design a network using only NAND gates which generates the logic function $\overline{\overline{A}.(B+\overline{C})}$.

3.5 The waveforms in figure 3.28a are applied to the circuit in figure 3.28b. Draw the waveform appearing at output f if its value is initially logic '0'.

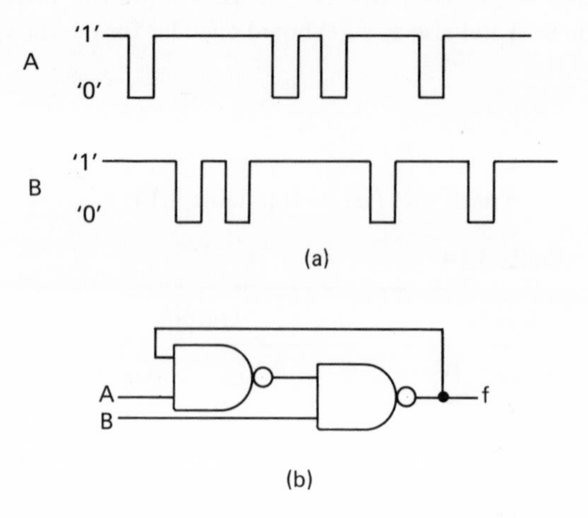

Figure 3.28

3.6 With the aid of a reasoned argument, give the addresses selected by the 1-out-of-8 decoder in figure 3.29 if the 'weights' of the 'select' inputs are X = 1, Y = 2 and Z = 4.

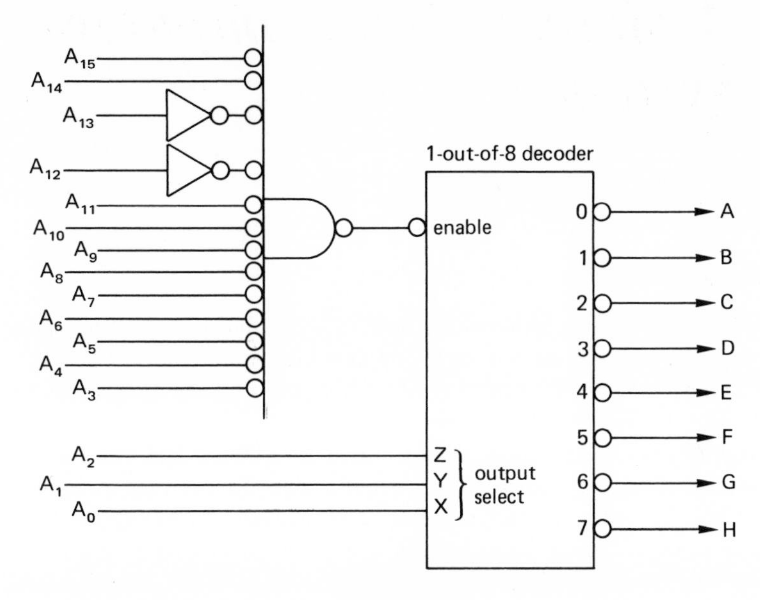

Figure 3.29

3.7 Show how two 8-bit priority encoders can be connected to a 16-key keyboard to provide a hexadecimal output.

4 A Simple Microcomputer System

4.1 The Bus System

As mentioned earlier, microprocessors operate on a 3-bus system. The majority of CPUs use a 16-line **address bus**, an 8-line **data bus**, and a **control bus** which contains a number of lines (the number of lines depends on the design of the CPU); we will return to the control bus later.

Microprocessors are generally manufactured in a 40-pin DIL form and, in the main, system considerations result in a requirement of more than 40 connections to the CPU. To overcome this problem, it is necessary to multiplex the signals applied to some of the CPU pins; that is, a number of the pins are used for a dual function. The functions which are multiplexed differ between manufacturers, and table 4.1 gives several examples. For example, in the Intel 8085 CPU the eight connection pins which function as the data bus at one moment, function as address bus lines $A_0 - A_7$ some time later.

Table 4.1

Manufacturer	Microprocessor	Multiplexed pins
Intel	8080	data bus/part of control bus
Intel	8085	data bus/$A_0 - A_7$
National	8060	D_0-D_3/A_{12}-A_{15} D_4-D_7/part of control bus

The multiplexing process permits the CPU to output certain information at one instant of time on some of the pins, and other information at another time on the same set of pins. If either set of information is required by the system, it is necessary to latch it in a support chip, otherwise the information is lost when the function of the pins changes.

For example, in an 8080 CPU system, vital control bus signals are output on the data bus at a certain point in time; to latch this data, it is necessary to connect the data bus lines to a support chip (an 8212 I/O chip is frequently used). When the 8080 CPU outputs the control bus signals on the data bus lines, it also provides synchronously a suitable strobe pulse; this strobe pulse is used to latch the control signals, which appear on the data bus lines, into the support chip. When the data bus lines are not used to supply the control bus signals, they are used for their main function, namely transmitting data both to and from the CPU.

In the National 8060 CPU, the four least significant data bus lines (D_0-D_3) are multiplexed with the four most significant address bus lines (A_{12}-A_{15}), and the four most significant data bus lines (D_4-D_7) are multiplexed with certain control signals.

It is clear from the above that not all of the lines of the three sets of buses are available to the user at the same time. However, in this book, we will assume that our CPU has sufficient connection pins to enable us to access all sixteen address bus lines and all eight data bus lines. Moreover, it is assumed that the vital control bus lines are also available.

A block diagram of the CPU is shown in figure 4.1; to simplify the diagram, a multi-wire bus is shown as a single wire with a stroke drawn through the wire, the number of wires in the bus being written by the side of the bus.

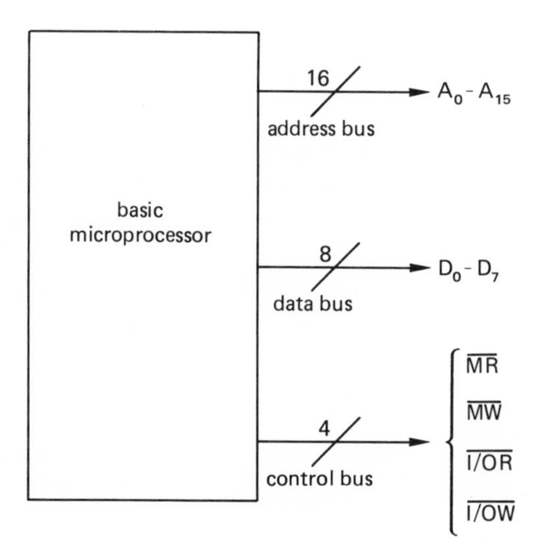

Figure 4.1

For the moment we will use a simple control bus, which contains only four wires, which are

$\overline{MR}$ (Memory Read - active low)
$\overline{MW}$ (Memory Write - active low)
$\overline{I/OR}$ (Input/Output Read - active low)
$\overline{I/OW}$ (Input/Output Write - active low)

When the program being executed calls for a memory read operation, the $\overline{MR}$

line is driven low, that is, it has a logic '0' on it, the other control bus lines remaining in the high state (logic '1'). On completion of the memory read instruction, the $\overline{MR}$ line returns to the logic '1' state. Similarly, when a memory write instruction is executed, the $\overline{MW}$ line is driven low; on completion of the instruction it returns to the high level. The waveform diagrams in figures 4.2 and 4.3 illustrate the operation during memory read and write operations and input/output operations, respectively.

The waveforms in the left-hand half of figure 4.2 are illustrative of those occurring during a typical **memory read** sequence. The CPU first presents the address of the memory to be selected on the address bus wires at point A. Since any one of the address wires may have either a logic '0' or a logic '1' on it (depending on the selected address), two lines are shown at point A in the figure - one corresponding to logic '0' and the other to logic '1'. For example, if the selected address is $2F09_{16}$, the logic level applied to address line A_0 is logic '1', that on line A_1 is '0' that on A_2 is '0', that on A_3 is '1' etc. At points where the address *may* change, such as at points A, E and I, the logic levels on the address bus wires are shown to cross.

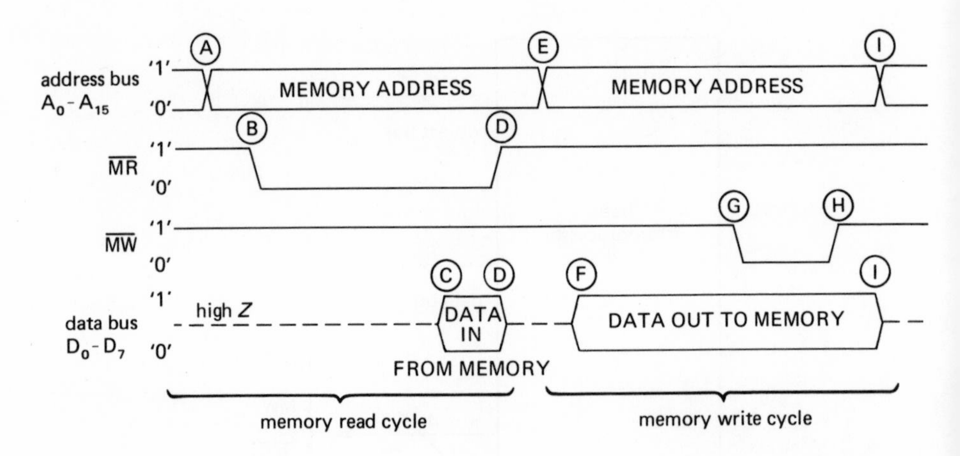

Figure 4.2

Shortly after the address of the memory has been established on the address bus, the $\overline{MR}$ (memory read strobe) is forced to logic '0' (point B); this 'enables' the memory element at the address specified by the signals on the address bus. It takes a little time before stable data are presented to the data bus lines, so that the data bus lines are maintained in their high impedance state until point C. Between C and D, the memory presents valid data to the CPU. At point D, the CPU causes the logic level on the $\overline{MR}$ to return to logic '1', by which time the data in the memory location has been read by the CPU. The program then changes the address at point E for the commencement of the next instruction which, in this case, is a memory write operation.

During a **memory write** cycle, the CPU presents the address to be selected to the address bus at point E (figure 4.2), and shortly afterwards it presents the data to be stored to the data bus at point F. At time G the CPU drives the memory write control bus line ($\overline{\mathrm{MW}}$) to logic '0'; this action results in the addressed location being 'selected'. After a short interval, the logic level on the $\overline{\mathrm{MW}}$ line is raised to logic '1', by which time data has been written into the selected location. At point I the address is changed once more under the control of the next instruction.

The operation of an **input cycle** is shown in the left-hand half of figure 4.3. In this case, the address of the input port is applied to the address bus at point I. A little time later the $\overline{\mathrm{I/OR}}$ control signal is forced to logic '0' (point J); this action enables the input port which is addressed by the signals on the address bus. Prior to the instant of time marked as K, all the lines on the data bus are maintained in their high impedance state; between K and L, the input port is given access to the data bus of the system, and valid data are transmitted to the CPU.

The sequence of events during an **output cycle** is shown in the right-hand half of figure 4.3. The signals giving the address of the selected output device (port) appear on the address bus at point M, and shortly afterwards (point N) the CPU outputs valid data to the data bus. When the $\overline{\mathrm{I/OW}}$ control signal is driven low (point P), this data is applied to the output device via the selected output port. When the data has been 'written' to the output port, the CPU forces the $\overline{\mathrm{I/OW}}$ signal high (point Q). After this, the signals on the address bus can be changed at point R under the control of the program.

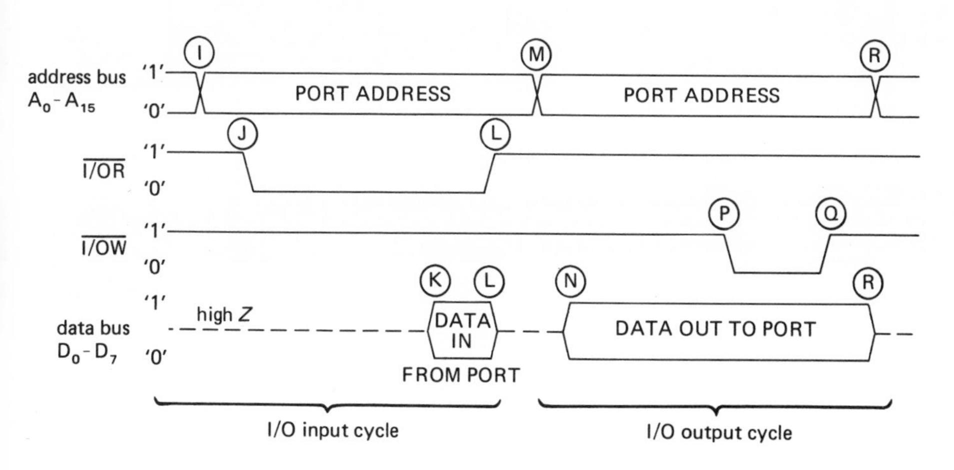

Figure 4.3

Although the four control signals used above, that is, $\overline{\mathrm{MR}}$, $\overline{\mathrm{MW}}$, $\overline{\mathrm{I/OR}}$ and $\overline{\mathrm{I/OW}}$ may not be directly available on all microprocessors, they can usually be derived from the control signals which are available. Each CPU has additional control signals, some of which are described later in the book.

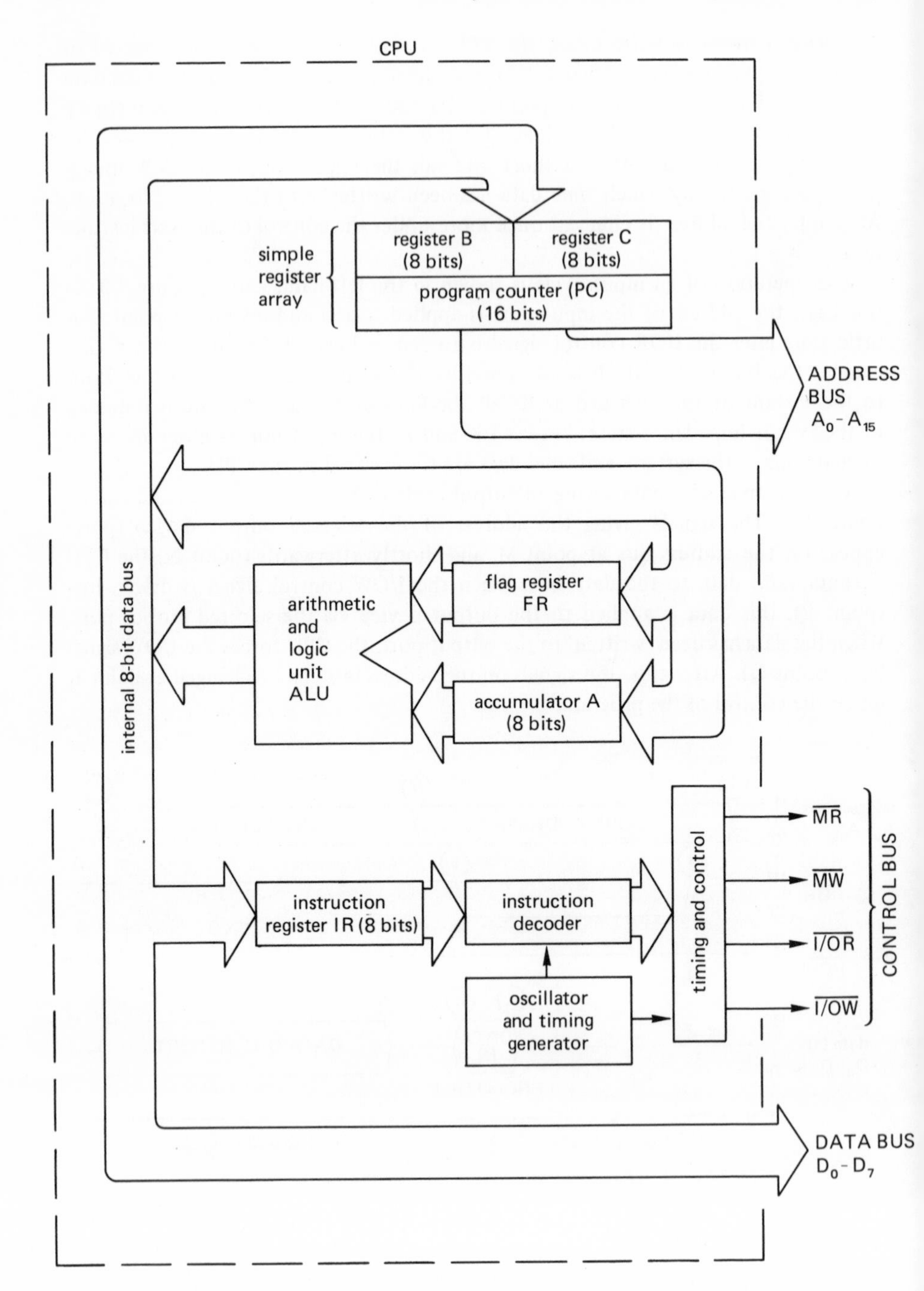
CPU
simple register array
register B (8 bits)
register C (8 bits)
program counter (PC) (16 bits)
ADDRESS BUS $A_0 - A_{15}$
internal 8-bit data bus
arithmetic and logic unit ALU
flag register FR
accumulator A (8 bits)
instruction register IR (8 bits)
instruction decoder
oscillator and timing generator
timing and control
$\overline{MR}$
$\overline{MW}$
$\overline{I/OR}$
$\overline{I/OW}$
CONTROL BUS
DATA BUS $D_0 - D_7$

Figure 4.4

4.2 Architecture of a Basic Microprocessor

For the purpose of simplicity, we consider here a microprocessor which has a limited number of registers (other registers are introduced later as the need arises). Some of the registers are **general-purpose registers** which the programmer can use to store either data or addresses. Additionally, the CPU also has a number of special-purpose registers used to store specific items of information. A block diagram illustrating the register arrangement and the **architecture** of the central processing unit (CPU) is given in figure 4.4; the architecture will be modified as the book unfolds to account for other aspects of the CPU. The registers involved in figure 4.4 are

the accumulator or A-register
the B-register
the C-register
the instruction register, IR
the flag register, FR
the program counter, PC

The A-, B- and C-registers together with the instruction register each store 8 bits, while the PC stores a 16-bit word. The flag register is an 8-bit register, 5 of these bits storing useful information while the remaining 3 bits store fixed logic levels (this arrangement is fixed by the CPU designer).

The **accumulator** is the heart of the microprocessor. The majority of the instructions executed by the microcomputer result in the accumulator being involved either in operations on data or in transfers of data. The **B-register** and the **C-register** can be used as temporary storage locations for data and other information; data can be moved between the A-, B- and C-registers under program control.

The **instruction register**, IR, is an 8-bit register which stores the binary instruction code currently being executed by the microprocessor. When the program calls for, say, a memory read instruction, the binary value corresponding to 'memory read' is transferred to the IR. This signal is decoded to cause, amongst other things, the $\overline{MR}$ line in the control bus to be driven low (logic '0'). This causes a memory element to be switched to its 'read' state.

As mentioned earlier, a **flag** is a flip-flop which is either set or cleared by a

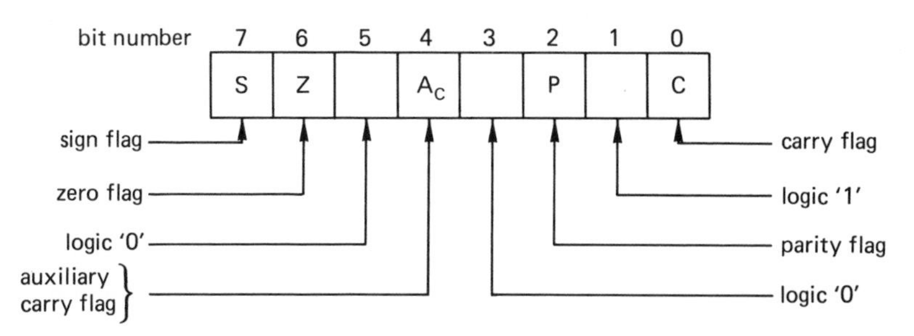

Figure 4.5

particular operation in the CPU. The **flag register** is a register containing five flags or flip-flops (it was mentioned above that the flag register is eight bits 'wide', but only five bits (the flags) are actively used). These flags are shown in figure 4.5; they are the **carry flag** (C), the **parity flag** (P), the **auxiliary carry flag** (A_C), the **zero flag** (Z) and the **sign flag** (S). The flags mentioned above refer to the 8080 microprocessor, and other manufacturers sometimes use flags for other purposes than those mentioned above. The state of the flags relates to certain operations involving the accumulator. The condition indicated by the state of each flag (that is, a '0' or a '1') is given in table 4.2.

Table 4.2

Flag	State	Meaning
C	0	A carry or borrow did not occur
	1	A carry or borrow did occur
P	0	The parity of the result is odd
	1	The parity of the result is even
A_C	0	The carry-out from bit 3 is '0'
	1	The carry-out from bit 3 is '1'
Z	0	The result is non-zero
	1	The result is zero
S	0	The sign of the result is positive
	1	The sign of the result is negative

To illustrate the effect of the execution of an instruction on the condition of the status flags, consider the outcome of adding 92_{16} to $F1_{16}$

$$F1_{16} = 11110001$$

$$92_{16} = \underline{10010010}$$

non-zero result; $Z = 0 \rightarrow$ 10000011 $\leftarrow$ three bits; $P = 0$

bit 7 = 1; S = 1 — no auxiliary carry; $A_C = 0$

carry occurs; C = 1

Arising from the above operation in the accumulator the state of the flags after the execution of the addition instruction are $C = 1$, $P = 0$, $A_C = 0$, $Z = 0$ and $S = 1$.

Programmers often refer to the 16-bit register pair combination comprising the contents of the 8-bit accumulator and the 8-bit flag word (which includes the three 'fixed' bits) as the **program status word** or **processor status word** (PSW).

The **program counter**, PC, is a 16-bit register containing the address of the next instruction to be executed by the CPU. The number stored in the PC is incremented during each instruction sequence; this ensures that the CPU 'knows' where the next instruction is to come from.

4.3 A Simple Microprocessor with I/O Facilities

In this section of the book, the reader is introduced to a microprocessor which is used to control a set of lamps via an output port (see figure 4.6). The operating program is stored in the ROM, and the lamps are connected to the data output lines of the output port. The signals which cause the lamps to be illuminated are produced either by the microprocessor itself (under the control of the program stored in ROM) or by the logic signals from the switches connected to the data input lines of the input port. A program used to control a set of lamps is described later in this chapter.

In the application described here, we do not need a RAM since the program can be stored in read-only memory (ROM); consequently the $\overline{\text{MW}}$ control signal (memory write strobe) is not needed in this system. The only lines needed in the control bus are therefore $\overline{\text{MR}}$ (memory read), $\overline{\text{I/OR}}$ (I/O read - which is used for reading data from the switches via the input port) and $\overline{\text{I/OW}}$ (I/O write - which is one of the signals used to 'enable' the output port).

ROM connections

The ROM in figure 4.6 stores 1024 bytes of data (a typical ROM of this kind is the 8308 chip), hence ten address lines (A_0 - A_9) are needed to address the individual storage locations (remember, $2^{10} = 1024$). In computer terminology, the value 1024 is referred to as lK; hence 2K bytes of storage correspond to 2048 8-bit words, 10K bytes correspond to 10 240 8-bit words, and 64K bytes to 65 536 8-bit words.

The ROM used in our microcomputer has two chip select lines ($\overline{\text{CS1}}$ and $\overline{\text{CS2}}$, both active low); $\overline{\text{CS1}}$ is energised from the remaining address lines (A_{10}-A_{15}) and $\overline{\text{CS2}}$ by the memory read strobe $\overline{\text{MR}}$. The lines in the address bus which are used to energise $\overline{\text{CS1}}$ are chosen by the system designer; in this case the system designer has decided to energise $\overline{\text{CS1}}$ from the output of an OR gate having address lines A_{10} to A_{15}, inclusive as its inputs. The reader will recall that the output from an OR gate is logic '0' only when all its inputs are logic '0' simultaneously; the $\overline{\text{CS1}}$ line of the ROM is therefore activated with a logic '0' only when the logic signals on the address bus lines A_{10} to A_{15} are simultaneously low.

The reader should note at this stage that the designer can choose alternative methods of providing a logic '0' signal on the $\overline{\text{CS1}}$ line; the circuit given here is simply one possible solution. Gate G1 may, in practice, not be a single 6-input OR gate but could be implemented by NAND gates or any other convenient logic elements. Moreover, it is not vital to use all six of the address lines (A_{10}-A_{15});

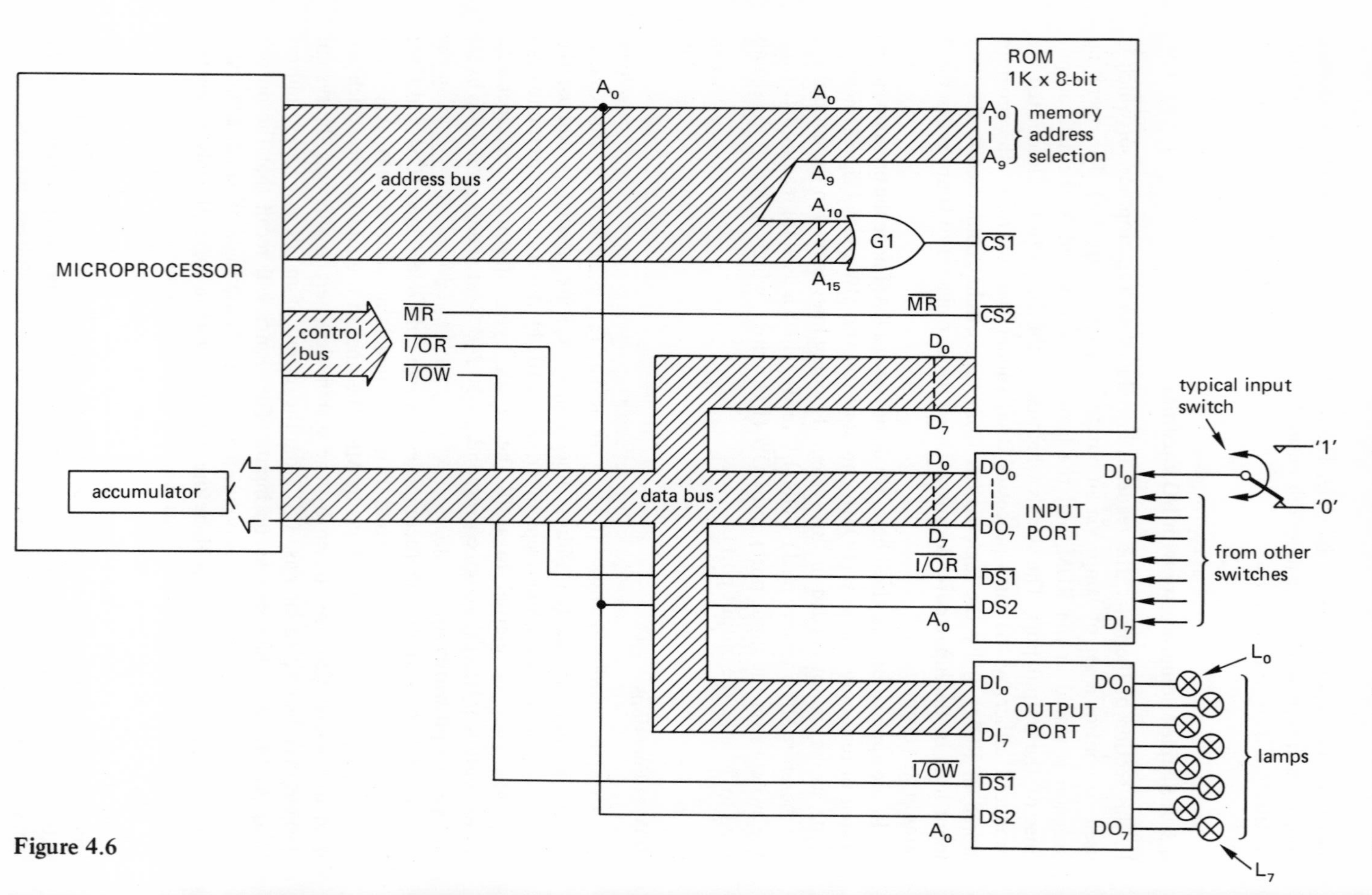
MICROPROCESSOR
accumulator
address bus
control bus
data bus
$\overline{MR}$
$\overline{I/OR}$
$\overline{I/OW}$
A_0
A_9
A_{10}
A_{15}
G1
ROM
1K x 8-bit
A_0 – A_9 memory address selection
$\overline{CS1}$
$\overline{CS2}$
D_0
D_7
INPUT PORT
DO_0
DO_7
DI_0
DI_7
$\overline{DS1}$
DS2
OUTPUT PORT
typical input switch
'1'
'0'
from other switches
lamps
L_0
L_7

Figure 4.6

many microcomputer designs use only one or two of these address lines for this application (see chapter 6 for details).

The $\overline{\text{CS2}}$ line of the ROM is energised by the $\overline{\text{MR}}$ strobe; that is, one condition for the ROM selection is that the CPU $\overline{\text{MR}}$ control bus line must be low.

The ROM in figure 4.6 is selected (enabled) when both $\overline{\text{CS1}}$ AND $\overline{\text{CS2}}$ are simultaneously low; this occurs when address lines A_{10} to A_{15} simultaneously have logic 0's on them AND when the $\overline{\text{MR}}$ is also low. When the ROM is enabled, the data at the address specified by the signals on address bus lines A_0-A_9, inclusive, is deposited on the data bus lines D_0-D_7, and is transmitted to one of the CPU registers.

Input port connections

The eight data input lines (DI_0-DI_7) of the input port are activated individually by logic signals from eight independent switches. The data output lines (DO_0-DO_7) of the input port are connected to the data bus lines (D_0-D_7) of the CPU. The input port in figure 4.6 is of the type described in chapter 3, and has two device select pins, namely $\overline{\text{DS1}}$ (active low) and DS2 (active high). For the input port to be selected, $\overline{\text{DS1}}$ must be low AND DS2 must simultaneously be high. Since the input device read strobe ($\overline{\text{I/OR}}$) in the control bus is low when an input device is being read, the $\overline{\text{I/OR}}$ control line is connected to the $\overline{\text{DS1}}$ line of the input port.

The DS2 line is activated by a signal from the address bus; to simplify the logic requirements of the system, only one address line (A_0) is used. Thus when a logic '1' appears on address line A_0, the DS2 pin of the input port is correctly energised for 'enabling' purposes. The state of the signals on the remaining address lines (A_1-A_{15}) does not matter (that is, any combination of 0's and 1's can occur on these lines) since they are not involved in the input port selection process.

Hence the input port is selected when $A_0 = 1$ AND a logic '0' appears on the $\overline{\text{I/OR}}$ control line. When this occurs, the logic signal from each switch is connected to the appropriate data bus line, and thence to the CPU.

Output port connections

The connections are similar to those of the input port, but in this case the CPU data bus lines D_0-D_7 are connected to the data input lines (DI_0-DI_7) of the output port. The data output lines of the port (DO_0-DO_7) are connected to eight lamps (L_0-L_7).

In the case of the output port, the $\overline{\text{DS1}}$ pin is activated by the $\overline{\text{I/OW}}$ control line signal and, as for the input port, the DS2 pin is connected to address bus line A_0. Hence the signals on the CPU data bus lines are applied to the lamps when the $\overline{\text{I/OW}}$ line is low AND the signal on the address bus line A_0 is high.

Summary

Table 4.3 summarises the state of the address lines and of the control bus lines for device selection to occur in figure 4.6.

Table 4.3

Device selected	Address bus line																Control bus line		
	A_{15}	A_{14}	A_{13}	A_{12}	A_{11}	A_{10}	A_9	A_8	A_7	A_6	A_5	A_4	A_3	A_2	A_1	A_0	$\overline{MR}$	$\overline{I/OR}$	$\overline{I/OW}$
ROM	0	0	0	0	0	0	X	X	X	X	X	X	X	X	X	X	0	X	X
Input port	X	X	X	X	X	X	X	X	X	X	X	X	X	X	X	1	X	0	X
Output port	X	X	X	X	X	X	X	X	X	X	X	X	X	X	X	1	X	X	0

X = 'Don't care', that is, could be '0' or '1'.

4.4 A Simple Program

In the following, a simple program is developed which causes the eight switches in the circuit in figure 4.6 to control eight lamps. The program outlined here is generally similar to that described at the end of chapter 1, the essential difference between the two is that we now deal with a more practical form of program.

Since the address bus is 16 bits 'wide', it is theoretically possible to address any one of 65 536 (64K) locations. That is, any address in the range 0000000000000000_2 (0000_{16}) to 1111111111111111_2 ($FFFF_{16}$) is available to the user. Since the microprocessor described here can only handle an 8-bit word, a problem immediately presents itself since a 16-bit address must be specified in two 8-bit sections as follows. The address bits which are output on the address bus lines A_7-A_0 are described as the low byte of the address, and address bits $A_{15}-A_8$ as the high byte; one method of expressing a 16-bit address is therefore in the form of two adjacent bytes in the program. This facility is very valuable when we need to enter (or re-enter) the program at a particular point. For example, if we need to enter the program at address $2F36_{16}$, the low byte of the address is stored in the program as 36_{16} ($0011\ 0110_2$) and the high byte of the address as $2F_{16}$ ($0010\ 1111_2$).

At the instant of switch-on, the microcomputer circuitry causes the contents of the program counter (PC) to be reset to zero (0000_{16}). As mentioned earlier, the PC is a register which contains the address of the next instruction to be executed by the CPU. The PC is often described as a **pointer register** which 'points' at a particular address. Since the PC is reset to zero at the instant of switch-on, the program therefore begins to run from address 0000_{16}.

The execution of an instruction can be divided into two broad sections, namely the instruction fetching sequence and the instruction execution sequence. During an **instruction fetching** sequence, the instruction word (see below) is fetched from the memory and is interpreted by the control unit; at the same time the value stored in the PC is **incremented**, so the PC points at the address of the next instruction to be fetched. During an **instruction execution** sequence, the microcomputer performs the operation called for by the instruction word.

When devising a program, it is often convenient to represent it initially in the form of a diagram which illustrates the sequence of events in the program. Such a diagram is known as a **flowchart**. Suppose, for example, that a part of a program is to call for data to be read or 'input' from the input port in figure 4.6, and then to output this data via an output port so that it illuminates a number of lamps. A flow chart associated with this small section of the program is illustrated in figure 4.7.

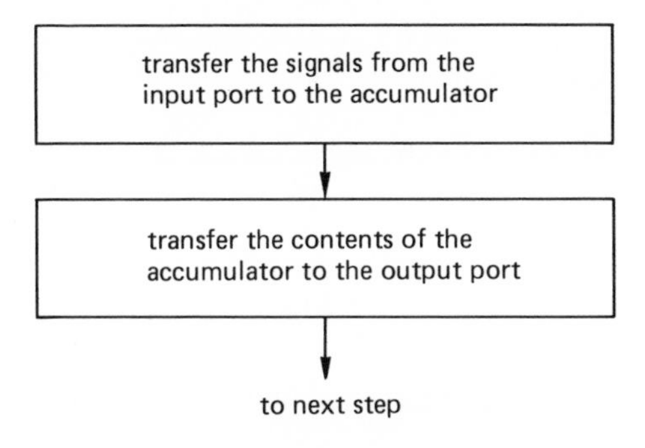

Figure 4.7

When writing a program associated with the flowchart in figure 4.7, the programmer must know the address of each of the I/O chips in figure 4.6. For the moment, the reader is asked to accept that the address of *both* I/O chips in figure 4.6 is 0001_{16} (methods of interpreting addresses from the electrical connections to chips are described in later chapters). In fact, with the instruction set adopted in this book, it is only necessary to specify the low byte of the address of I/O ports. Thus both I/O chips in figure 4.6 have the address 01_{16}.

The program associated with the flowchart in figure 4.7 may be written as follows:

INput the contents of input port 01 to the accumulator.
OUTput the contents of the accumulator to output port 01.

To reduce the effort of writing the above two lines out in full, computer users adopt a simplified language known as an **assembly language**. Each instruction in this language is written in the form of a mnemonic which defines the operation to be carried out. Thus the mnemonic IN means 'input the contents of a specified port to the accumulator', and OUT means 'output the contents of the accumulator to a specified port'. Thus the program relating to figure 4.7 can be written as

Instruction	Comment
IN 01	input data from switches
OUT 01	output data to lamps

The above section of program is a simplified form of what is known as a **source program**; a source program cannot be directly executed by a microprocessor, and requires translating into an **object program**. The translation process is known as **assembly**. The assembled object program contains information which allows the program to be run by the microprocessor; an example of the additional information in the object program is the address of each instruction. Each instruction mnemonic or **source statement** is converted into a binary **object code** (known also as an **opcode**) by the assembler. For example, the object code for the INput instruction is DB_{16} (11011011_2), and that for the OUTput instruction is $D3_{16}$ (11010011_2). Thus the above instructions appear as follows

Memory address (hex)	Memory contents (hex)	Instruction	Comment
0000	DB	IN 01	INPUT data
0001	01		
0002	D3	OUT 01	OUTPUT data
0003	01		

Now we look more closely at the operation of the microcomputer during IN and OUT instructions.

INput instruction

Figure 4.8 shows the register elements involved in the execution of an INput instruction; these are the instruction register (IR), the program counter (PC) and the accumulator (A).

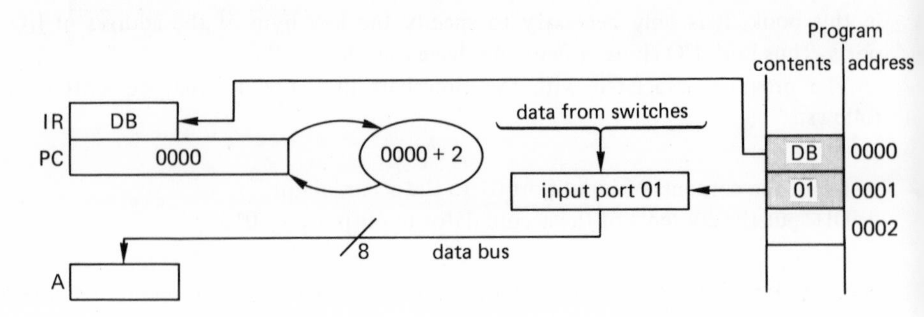

Figure 4.8

The control unit of the CPU first transfers the contents of program address 0000 to the IR, at which point the CPU interprets it as a two-byte instruction. The

PC is therefore incremented by two, so that the value it stores becomes 0002_{16}; this address contains the next instruction to be fetched by the CPU. The control unit recognises the contents of the IR (that is, DB_{16}) as an INput instruction, and it identifies the contents of the second byte of the IN instruction as the address of the input port from which data is to be fetched. Consequently, address 01_{16} is output on the address bus and, at much the same time, the $\overline{I/OR}$ control bus is forced low by the CPU. The combined effect of this action has two results; firstly, it enables the input port 01 and, secondly, it inputs the state of the eight switches to the accumulator via the data bus. The input data is applied to the CPU for a few hundred nanoseconds only, but this is sufficient time for the information to be latched into the accumulator, where it remains after the input port is finally disabled. Thus if a logic '1' signal is applied to the DI_0 line of the input port, and logic '0' to the DI_7 line, then after the IN instruction a '1' is stored in bit 0 of the accumulator and a '0' is stored in bit 7.

The process of writing data into the accumulator during the INput sequence destroys the data originally stored in the accumulator.

OUTput instruction

The movement of data between the accumulator and the output port during an OUTput instruction is shown in figure 4.9. The two byte OUTput instruction is stored at addresses 0002 and 0003. During the instruction fetch sequence, the PC is incremented by two, so that it stores 0004, which is the address of the next instruction to be fetched. During the execution stage of the OUTput instruction,

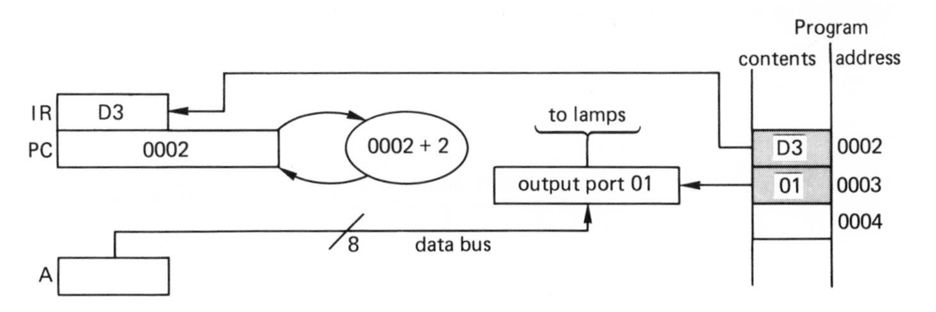

Figure 4.9

the CPU outputs address 01_{16} on the address bus AND a logic '0' on the $\overline{I/OW}$ control line. This allows the data in the accumulator to be applied to the lamps connected to the output port. If logic '1' and logic '0', respectively, are stored in bit 0 and bit 7 of the accumulator, then lamp L_0 (see figure 4.6) is illuminated and L_7 is extinguished.

The process of reading data from the accumulator is non-destructive; that is, the data in the accumulator is unchanged after the OUTput sequence. The data in the accumulator is changed only when new (changed) data is transferred into the accumulator.

Conclusion

In the above, the reader has been introduced to the method used to control the flow of data through the microcomputer. However, an important point remains unresolved, namely what happens next? In general, the programmer must ensure that the CPU has 'something to do'. It is therefore necessary for the programmer to provide a valid instruction at memory address 0004 to ensure orderly operation. If the programmer has not done this, the CPU will interpret the group of 1's and 0's in location 0004 (which is electronic 'garbage') as an instruction and will attempt to execute it; this may lead to disastrous results.

A possible solution to the problem in this case is to instruct the CPU to return to the beginning of the program at the completion of each INput/OUTput sequence. That is, an unconditional jump must be made from the end of the sequence to address 0000_{16}. This is one of many versions of the **jump instruction** which are available in microprocessors. In this way the state of the switches is continuously scanned, the switches apparently controlling the lamps as though each switch were connected to a single lamp. This modification is dealt with in section 4.5.

4.5 The Unconditional Jump (JMP) Instruction

The JuMP (JMP) instruction is one which allows the CPU to JuMP unconditionally from one point in the program to another point; it is used here to cause the program described in section 4.4 to return to its beginning. This is illustrated in the flow-chart in figure 4.10, the operations within the registers being shown in figure 4.11.

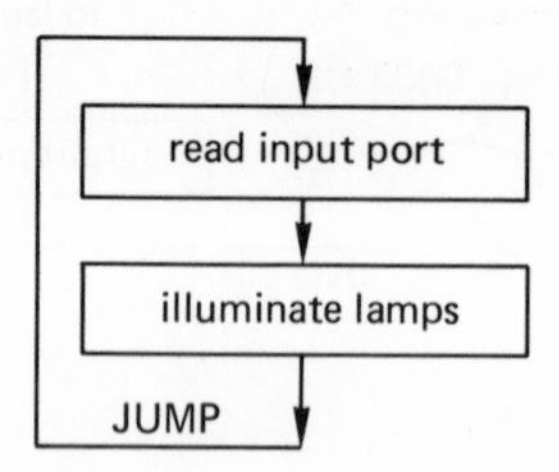

Figure 4.10

The JuMP instruction defines all sixteen bits of the address to which the JuMP is made; it is therefore a 3-byte instruction occupying program addresses 0004, 0005 and 0006. Address 0004 contains the hex object code (C3) representing the JMP instruction, the second byte (in address 0005) contains the low byte of the required address, and the third byte (in address 0006) contains the high byte of the destination address. Before the JMP instruction is executed, the PC stores

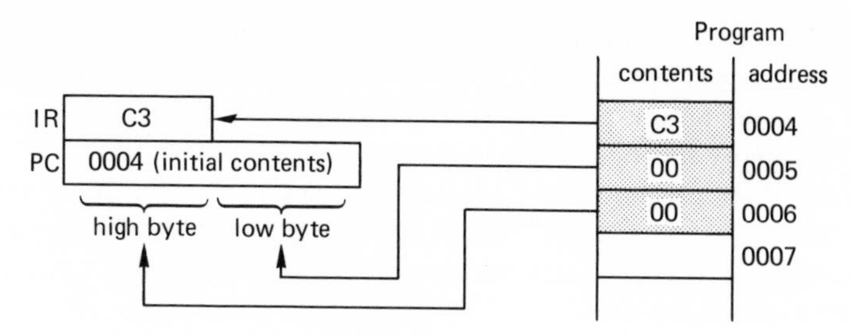

Figure 4.11

0004; on completion of the instruction it stores 0000, the initial contents of the PC being 'lost'. The program may therefore be written as follows

Address (hex)	Machine code	Instruction	Comment
0000	DB	IN 01	INPUT data from switches
0001	01		
0002	D3	OUT 01	OUTPUT data to lamps
0003	01		
0004	C3	JMP 0000	JUMP to beginning
0005	00		
0006	00		

The JMP instruction does not alter the contents of the accumulator. If we wish to make an unconditional jump to, say, address $OF23_{16}$, then instructions 0004 - 0006 would read

Address (hex)	Machine code	Comment
0004	C3	
0005	23	low byte of address
0006	OF	high byte of address

4.6 Control Bus Signals

The control bus lines $\overline{MR}$, $\overline{MW}$, $\overline{I/OR}$ and $\overline{I/OW}$ used throughout this book may not be available on all CPU chips. It may therefore be necessary to decode these signals from others which are available on the CPU chip. The CPU will have a read strobe

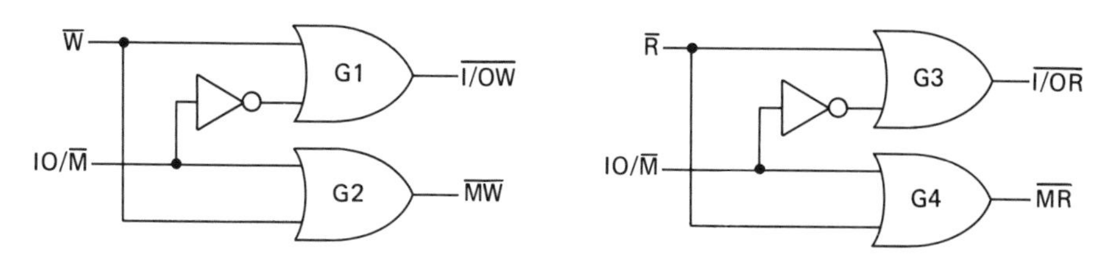

Figure 4.12

($\overline{READ}$ or $\overline{R}$, active low) and a write strobe ($\overline{WRITE}$ or $\overline{W}$, active low). To generate the signals mentioned in the first sentence of this section, the user needs to have access to an input-output/memory strobe ($IO/\overline{M}$, active high for I/O operations and low for memory functions).

One method of generating the control bus signals using the $\overline{W}$, $\overline{R}$ and $IO/\overline{M}$ signals is shown in figure 4.12. The CPU generates appropriate $\overline{W}$, $\overline{R}$ and $IO/\overline{M}$ signals in response to instructions it executes. The output signals from the circuit in figure 4.12 are listed below.

Inputs		Outputs		Inputs		Outputs	
$\overline{W}$	$IO/\overline{M}$	$\overline{I/OW}$	$\overline{MW}$	$\overline{R}$	$IO/\overline{M}$	$\overline{I/OR}$	$\overline{MR}$
0	0	1	0	0	0	1	0
0	1	0	1	0	1	0	1
1	0	1	1	1	0	1	1
1	1	1	1	1	1	1	1

PROBLEMS

4.1 For the circuit in figure 4.13 state (a) if the port is an input port or an output port, (b) the control signal which has to be connected to the $\overline{DS1}$ pin (that is, $\overline{I/OR}$ or $\overline{I/OW}$) and (c) the address of the port.

4.2 Based on a microprocessor of your own experience, describe how the control signals $\overline{MR}$, $\overline{MW}$, $\overline{I/OR}$ and $\overline{I/OW}$ are generated from the signals produced by the CPU.

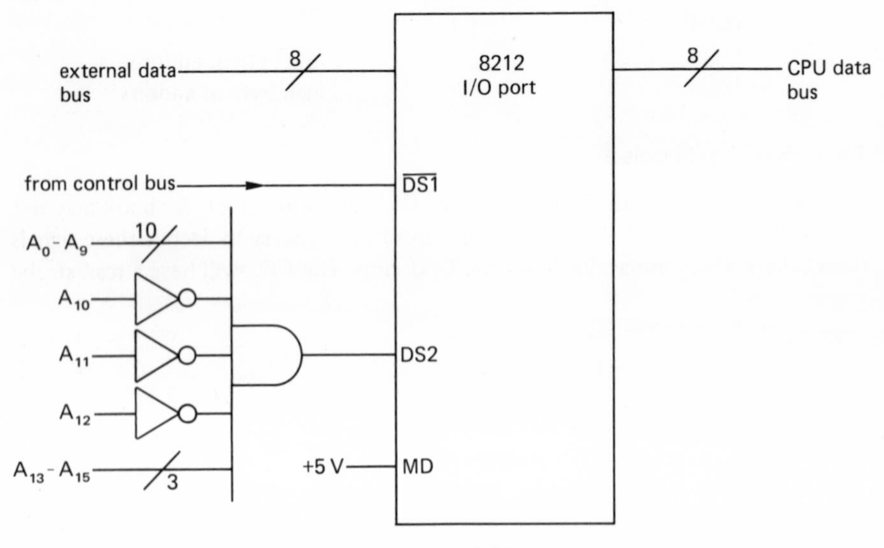

Figure 4.13

5 Requirements of a Microcomputer System

5.1 A Practical System

In this chapter we discuss the general requirements of a practical system. Figure 5.1 shows a block diagram of a functional microcomputer system. The system lacks only a few components to make it into a practical working system; these components include an address bus driver and a data bus driver, which ensure that the address bus and the data bus, respectively, can deal with the loads connected to them. These components have been omitted since they are not relevant to a description of a basic system. The support chips shown in figure 5.1 are

(1) a 1K x 8-bit read-only memory (ROM)
(2) a $\frac{1}{4}$K x 8-bit random-access memory (RAM)
(3) two input ports
(4) two output ports

Many microcomputers also require a chip known as a **system controller** (although other names are also used for this chip). This chip takes the control signals produced by the CPU and converts them into the control signals used in the control bus (which in our case are $\overline{MR}$, $\overline{MW}$, $\overline{I/OR}$ and $\overline{I/OW}$). We have dispensed with the system controller in figure 5.1 by the simple device of assuming that the correct control signals are generated by the CPU.

ROM

This contains the permanent program used in the microcomputer. The program could relate, for example, to a traffic light control scheme in which signals from sensors in the road are transmitted to the CPU via an input port, and are used to control the traffic lights which operate from signals generated by the CPU.

The signal applied to the $\overline{CS1}$ pin of the ROM chip is generated from address bus lines $A_{10}-A_{15}$, inclusive, via an OR gate; thus each of these address lines must have a logic '0' on them to drive the CS1 line low. The $\overline{CS2}$ pin of the ROM is activated by the $\overline{MR}$ control line. To select the ROM, both $\overline{CS1}$ and $\overline{CS2}$ must be low simultaneously; the memory location within the ROM which is identified by the signals on address lines A_0-A_9, inclusive, is then connected to the data bus.

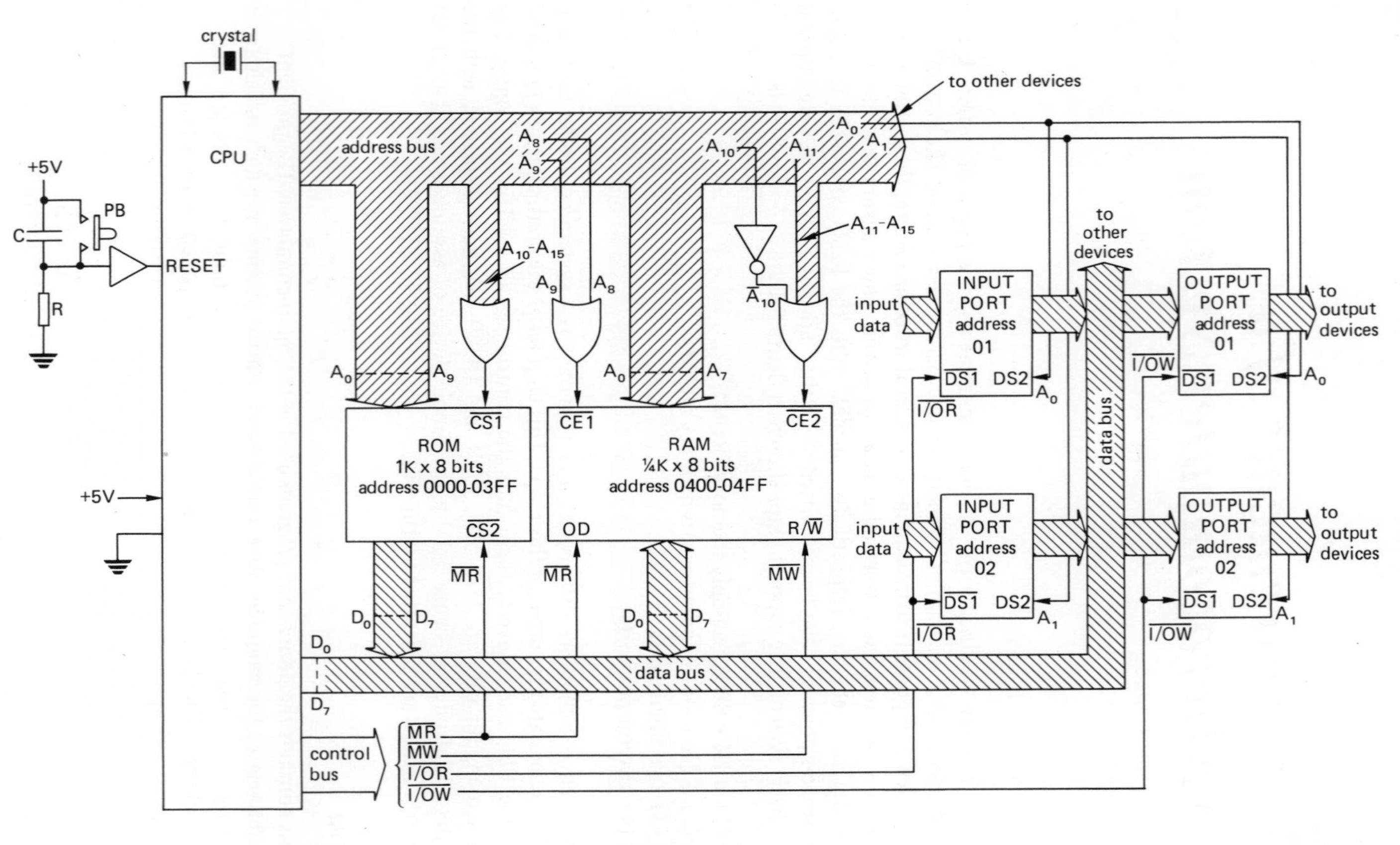
crystal
CPU
+5V
PB
C
RESET
R
+5V
address bus
to other devices
A_0
A_1
A_8
A_9
A_{10}
A_{11}
A_{10}–A_{15}
A_{11}–A_{15}
$\overline{A_{10}}$
A_0 A_9
A_0 A_7
$\overline{CS1}$
ROM
1K x 8 bits
address 0000-03FF
$\overline{CS2}$
$\overline{CE1}$
$\overline{CE2}$
RAM
¼K x 8 bits
address 0400-04FF
OD
R/$\overline{W}$
$\overline{MR}$
$\overline{MW}$
D_0 D_7
data bus
control bus
$\overline{MR}$
$\overline{MW}$
$\overline{I/OR}$
$\overline{I/OW}$
input data
INPUT PORT address 01
INPUT PORT address 02
$\overline{DS1}$ DS2
to other devices
OUTPUT PORT address 01
OUTPUT PORT address 02
to output devices

Figure 5.1

Input port

The signals applied to the data input (DI) lines of input port 01 are connected to the data bus when A_0 is logic '1' AND $\overline{I/OR}$ is logic '0'. The data applied to input port 02 is read when A_1 is high AND $\overline{I/OR}$ is low simultaneously.

Output port

The signals on the data bus are applied to the devices connected to output port 01 when A_0 is logic '1' AND $\overline{I/OW}$ is logic '0'. Data is written to output port 02 when A_1 is high AND $\overline{I/OW}$ is low.

RAM

The **random-access memory** (RAM) is a read-write memory which allows the user either to write data into a particular location (address), or to read data from an address. The RAM is an area of memory which is used to store data temporarily. For example, if the microcomputer carries out a traffic census in addition to its main function (which may be that of a traffic light controller), it must have an area of memory in which to store details of the number of vehicles passing in each direction across the junction. This data must be stored in RAM, since the value stored must be changed (or updated) every time a vehicle passes across the junction. The operation of random access memories is dealt with in detail in chapter 6.

The chip enable lines shown in figure 5.1 are typical of a small RAM, the chip being enabled when $\overline{CE1}$ AND $\overline{CE2}$ are both low simultaneously. Line $\overline{CE1}$ has a logic '0' applied to it when address lines A_8 AND A_9 have logic 0's on them simultaneously. Line $\overline{CE2}$ is low when address line A_{10} is high and lines A_{11} to A_{15}, inclusive, are low simultaneously. The RAM chip is therefore enabled when the following conditions apply

	A_{15}	A_{14}	A_{13}	A_{12}	A_{11}	A_{10}	A_9	A_8	A_7	A_6	A_5	A_4	A_3	A_2	A_1	A_0
binary	0	0	0	0	0	1	0	0	X	X	X	X	X	X	X	X
hex	0				4				0 - F				0 - F			

The 'X' values in the above could either be '0' or '1'; the reason for this is that address lines A_7 to A_0 are used to select the $2^8 = 256$ ($\frac{1}{4}$K) storage locations in the RAM. The address of a location in the RAM therefore lies in the range 0400_{16} to $04FF_{16}$.

Since data can either be read from or written into the RAM, the data link (D_0 - D_7) must be bidirectional; this is indicated by the arrows pointing in opposing directions on the link between the RAM and the data bus in figure 5.1. However, the system must ensure that the CPU does not attempt to write data into the RAM and, simultaneously, to read data from it. To make sure that the CPU either only writes into the RAM or, alternatively, reads from it, the chip has two control lines (OD and $R/\overline{W}$) on it. The OD line is an **output disable** control line which, when it has a logic '1' applied to it, prevents data from being read from the RAM; for this reason the OD line is energised by the $\overline{MR}$ control signal.

When the CPU needs to read data from the RAM it forces the $\overline{MR}$ line low, allowing data to be read from the RAM. Additionally, the $\overline{MW}$ control signal is connected to the $R/\overline{W}$ line (Read/NOT Write); data are read from the memory when this signal is high, and data can be written into memory when this line is low. Hence, when the $\overline{MW}$ control signal is logic '0', data can be written into the RAM.

Additional connection to the CPU

Accurate timing of events is vital in microprocessor systems, and for this reason the majority of CPUs use a crystal-controlled oscillator as the timing element. The crystal is external to the CPU as shown in figure 5.1.

Additionally, it is sometimes necessary to reset the address of the program counter to 0000_{16}, that is, to cause the program to start (or restart) at the beginning. This can be brought about by means of a signal generated by a hardware circuit. In figure 5.1, this signal is developed by means of the circuit connected to the RESET input line of the CPU, and consists of an R–C circuit, a push-button and a buffer amplifier. When power is first applied to the CPU, capacitor C charges from the 5 V supply via resistor R; initially the charging current is high, so that the voltage across R is about 5 V. This voltage is applied to the RESET (active high) line of the CPU via the buffer amplifier; this logic '1' signal causes the program counter to 'point' to address 0000_{16}.

Once the capacitor has charged fully, the charging current falls to zero; at this point the voltage across R is also zero, allowing the CPU to proceed with the stored program. If the push-button (PB) is pressed at any time during the normal running of the program, it discharges capacitor C and momentarily applies a logic '1' to the RESET line. This action forces the program to return to address 0000_{16}.

Summary

The block diagram of the microcomputer system in figure 5.1 may correspond either to a dedicated system or to a general-purpose system. A **dedicated micro-computer** is one which is dedicated to the solution of a particular problem, such as the control of a set of traffic lights at a road junction. In this case, the input signals are derived from sensors in the road, and output signals are applied to the traffic lights. In a **general-purpose system**, the input signals would, for example, be obtained from a keyboard and the output signals would operate display devices such as seven-segment displays or a video display.

In the case of a dedicated system the majority of the program is stored in ROM; unless the program calls for extensive data manipulation, the storage requirement of the RAM will be minimal (see also section 6.13).

In a general-purpose system, the user requires the microcomputer to perform a number of widely differing programs, and it is likely that the RAM storage requirement will be fairly extensive. In this case, one of the main functions of the program stored in the ROM is to scan the state of the input devices, one of which may be a keyboard, and also to output information on a display device.

5.2 Monitor Program

A **monitor program** gives the microcomputer user assistance in running and developing other programs. In addition, the monitor usually provides additional software resources, such as subroutines which can be 'called up' to activate displays on output devices, etc.

The monitor program is usually stored in ROM, and provides the user with the ability to load programs from input devices, to read the contents of memory locations, to start programs from any location, and to run a program for a given number of instructions. The monitor is not usually very large (say $\frac{1}{2}$K byte to 2K byte in length) and is generally easy to understand; for this reason the user is advised to study the monitor program, since it provides some information about the use of the programming language. The minimal features of any monitor program are

(1) to provide control of the programmers input terminal;
(2) to store the program in successive bytes of memory;
(3) to allow the user to examine the contents of storage locations;
(4) to allow the user to load the 'start' address of the program so that control can be transferred to the program.

The monitor often provides other features including the ability to load data from and to store data in external storage devices such as magnetic tapes and discs.

As mentioned above, the monitor may reside in ROM while the CPU executes other programs. In this case it is known as a **resident monitor**. In some cases the monitor can be loaded into RAM (a **loadable monitor**), in which case some means must be provided for entry into the monitor program (such as an **interrupt** signal - see chapter 9).

5.3 Keyboard Scanning

Keyboards are used not only for the manual control of microcomputers, but also for manually entering data into storage locations. The basic circuit associated with one key of a keyboard is shown in figure 5.2. The contacts are of the 'normally open' type, so that the output signal is logic '1' in the 'normal' (unoperated) state. When the key is depressed, the output line is earthed and the output signal is logic '0'.

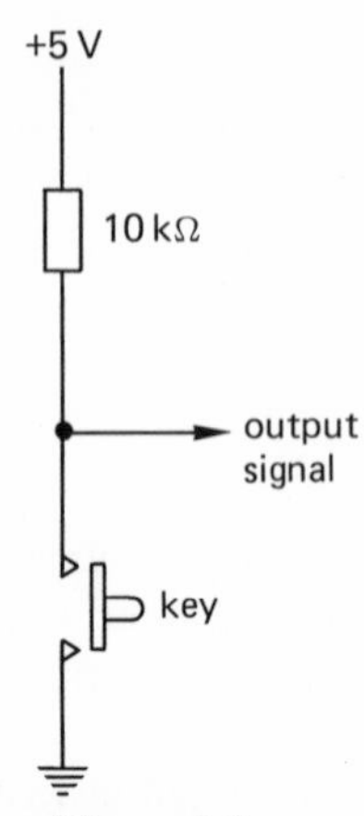

Figure 5.2

The monitor program causes the CPU to scan the complete keyboard periodically. The keys normally used are electromechanical devices, and display a phenomenon known as **contact bounce**. Contact bounce arises from the spring-inertia combination of the contact mechanism, and results in the contacts bouncing when the key is either pressed or released. Various methods are used to overcome the effects of contact bounce, one being to ensure that the keyboard is scanned or is serviced at intervals of not less than 5–10 ms. This period of time generally exceeds the maximum contact bounce time of the keyswitch (note: in a few cases, a time longer than 10 ms is necessary).

A popular type of keyboard scanning system is shown in figure 5.3. When a key is pressed, its contacts connect a column wire to a row wire; for example, K_3 serves to connect column wire C_0 to row wire R_1. The keyboard is connected to the data bus by an input port (whose address is 01 – see figure 5.1 as an example of a typical system) and also by means of an output port (whose address is also 01).

The keyboard described here is a simple 3 x 3 matrix containing nine key-switches K_0 to K_8. The keyboard is scanned by signals applied to column wires C_0 to C_3, inclusive, and the condition of the switches is indicated by the logic level on row wires R_0 to R_3. Each row wire is connected to a +5 V supply via a 10 kΩ current-limiting resistor, so that in the absence of any key being pressed, each row wire has a logic '1' on it. The complete keyboard scanning process usually involves two steps as follows:

step 1: check for key closure (that is, any one of the nine keys)
step 2: identify the key which is closed

Step 1 is a relatively simple procedure which can be carried out quickly, while step 2 takes a little longer. For this reason, the second step is completed only when the CPU is aware that a key has been pressed. Otherwise, the CPU continues with

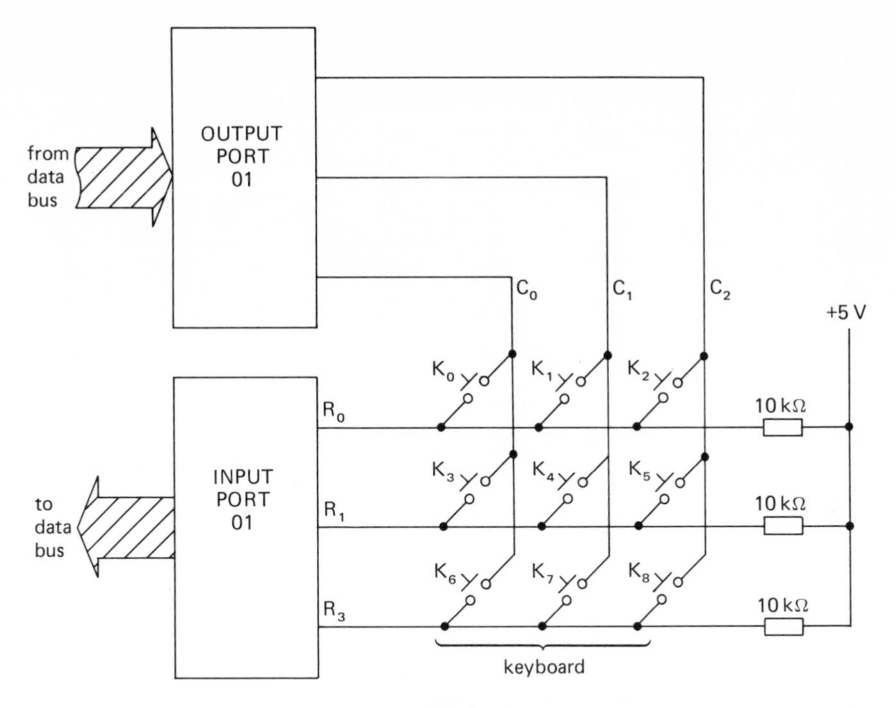

Figure 5.3

other sections of the program after step 1 indicates that none of the keys has been pressed. In the description which follows, it is assumed that column wire C_0 is connected via the output port to data bus D_0, and that row wire R_0 is connected via the input port to data bus D_0. Similarly, C_1 and R_1 are connected to data bus wire D_1 via the appropriate port, etc.

Step 1: check for key closure

A flowchart for a routine which checks if any key has been pressed is shown in figure 5.4. As mentioned earlier, flowcharting is a method widely used to represent a sequence of events in a computer program, and is represented by an interconnected set of symbolic blocks. For example, the rectangular symbol in figure 5.4 represents a general operation in the program. The diamond-shaped symbol in figure 5.4 is a decision symbol, in which a decision is required on the statement within the symbol. Depending on the result, the program follows either the NO path or the YES path from the symbol.

The flowchart in figure 5.4 does not allow for the contact to be debounced, but a minor modification within the 'ground keyboard columns' box allows debouncing to be carried out. The general procedure is to apply logic 0's simultaneously to each column wire (C_0, C_1, C_2) via output port 01 (see figure 5.3); this corresponds to the 'ground keyboard columns' operation in the flowchart. The data

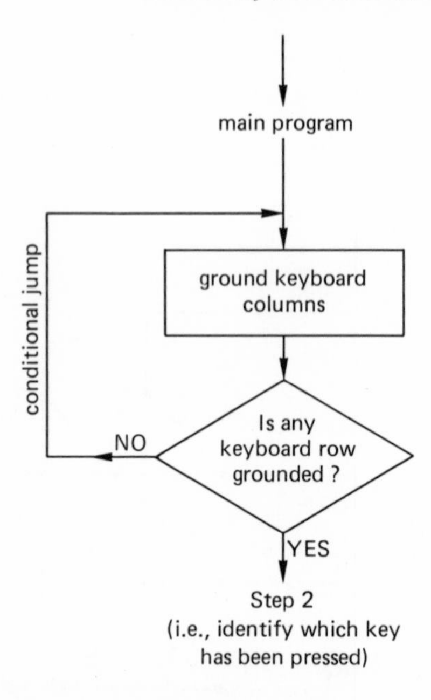

Figure 5.4

returned to the data bus via input port 01 is then checked by the CPU to determine if any row wire has a logic '0' on it; this corresponds to the 'is any keyboard row grounded' decision symbol. If each row wire has a logic '1' on it, then none of the keys has been pressed; in the program considered here, this results in an unconditional jump to be made via the 'NO' path in the flowchart to the point in the program where the column wires are grounded again. The net result is that the CPU executes a **waiting loop** until a key is pressed.

When **any key** is pressed, one of the row wires has a logic '0' applied to it; the microprocessor quickly detects this condition and allows the program to continue to step 2; that is, it identifies the key which has been pressed. A program satisfying the flowchart in figure 5.4 is given in table 5.1.

The program in table 5.1 introduces a number of new instructions, which are described below; since the start address of the program is 0030, we may assume that all instructions up to and including 002F have already been executed. The machine code 3E stored in location 0030 is one of the **M**o**V**e **I**mmediate (MVI) instructions the CPU can execute. In this case, 3E causes the control unit to MoVe the value specified in the second byte (that is 00_{16}) Immediately into the accumulator; the full range of MoVe instructions are described in chapter 10. On the completion of this instruction, the accumulator stores 00_{16} (00000000_2).

Next, the instruction D3 (in location 0032) causes the CPU to OUTput the contents of the accumulator to the output port at address 01 (the reader should also refer to figure 5.3). This instruction grounds the column rows of the keyboard.

Table 5.1 A simple keyboard scanning routine

Address (hex)	Machine code	Instruction mnemonic	Comment
0030	3E	MVI 00_{16}	Move 00000000_2 into accumulator
0031	00		
0032	D3	OUT 01	Ground columns
0033	01		
0034	DB	IN 01	Read state of the keyboard
0035	01		
0036	E6	ANI 00000111_2	Mask row data
0037	07		
0038	FE	CPI 00000111_2	Has a key been pressed?
0039	07		
003A	CA	JZ 0030_{16}	If not, return to 0030_{16}
003B	30		
003C	00		
003D			Step 2 of program
003E			

In fact, the same result is achieved by outputting any binary word containing 0's in the three least significant positions of the word; thus the data at address 0031 in the program may be 08, or 10, or 18, . . . F8.

The next two instruction bytes, DB 01 (in locations 0034 and 0035), cause the eight bits associated with input port 01 to be INput to the accumulator (also refer to figure 5.3). The reader is reminded that although we know that row wire R_0 of the keyboard is connected to data bus D_0, that row wire R_1 is connected to D_1, and that row wire R_2 is connected to D_2, we do not know what connections are made to the remaining terminals of the input port. To economise on the amount of hardware needed in the microcomputer, it is probable that the remaining terminals of the input port have signals on them from other sensors or transducers; since these signals are irrelevant to the problem in hand, we need to **mask** the signal from input port 01 so that the CPU 'sees' only the relevant information from the keyboard rows.

The masking process is carried out by the 2-byte instruction in locations 0036 and 0037. The first byte of the instruction (machine code E6 at address 0036) calls for the CPU to **AND** the contents of the accumulator **I**mmediately with the data in the second byte of the instruction, that is, AND the accumulator contents with 07_{16} or $0000\ 0111_2$. Suppose that after the 2-byte INput instruction has been executed the accumulator contains AF_{16} or 10101111_2 (note: the five most significant bits are presumed to come from sensors connected to input port 01), the effect of the ANI 07_{16} is as follows.

Initial contents of the accumulator	10101111
07_{16}	00000111
Final contents of the accumulator	00000111

From earlier work, the reader will appreciate that each pair of aligned bits in the above example is ANDed together to give the result. The net result of the ANI 07 instruction is therefore to 'mask out' irrelevant data in the input word, leaving the accumulator with only the relevant information from the keyboard.

Having masked out the irrelevant data, it is necessary to check if any of the keys have been pressed, that is, are any of the three l.s.b.s in the accumulator zeros? If none of the keys has been pressed, the accumulator contains 07_{16} after the completion of the ANI instruction; if one or more keys have been pressed, then it contains a value in the range 00_{16} to 06_{16}. For example, if one of the keys K_0, K_1 or K_2 (see figure 5.3) has been pressed, it grounds row R_0, resulting in $R_0 = 0$, $R_1 = 1$, $R_2 = 1$. After the ANI 07 instruction the accumulator contains 06_{16}, as will be seen from the following

	random data / data from keyboard rows
Initial contents of the accumulator	10101110
07_{16}	00000111
Final contents of the accumulator after ANI instruction	00000110 (0_{16} 6_{16})

A simple method of checking if the accumulator 07_{16} (that is, keys not pressed) is to subtract 07_{16} from the contents of the accumulator, and then to test the resulting contents of the accumulator to see if it is zero. If the result of the test shows that the accumulator stores zero, then none of the keys has been pressed; in this case, the CPU must jump back to the commencement of the sequence to test the keyboard once more (see figure 5.4). However, if the result of the test is nonzero, then one of the keys has been pressed, and step 2 of the program can be executed.

The 'subtract 07_{16}' operation suggested above is replaced here by a **CPI** instruction (ComPare Immediate the contents of the accumulator with data specified in the program). The comparison operation is generally similar to that of subtraction, but the result is not registered in the accumulator as a numerical value; in fact, the result merely alters the status flags. Referring to the program in table 5.1, the reader will note that following the ANI instruction, the CPU executes a 2-byte CPI instruction. This calls for the CPU to ComPare Immediately the contents of the accumulator with the hex value 07 in the second byte of the instruction. The movement of data during the CPI instruction is shown in figure 5.5. If the accumulator contains 06_{16} after the ANI instruction (as it will if one of the keys K_0, K_1 or K_2 has been pressed), the CPI 07 instruction causes 07_{16} to be subtracted from 06_{16}, the result of the operation simply altering the contents of

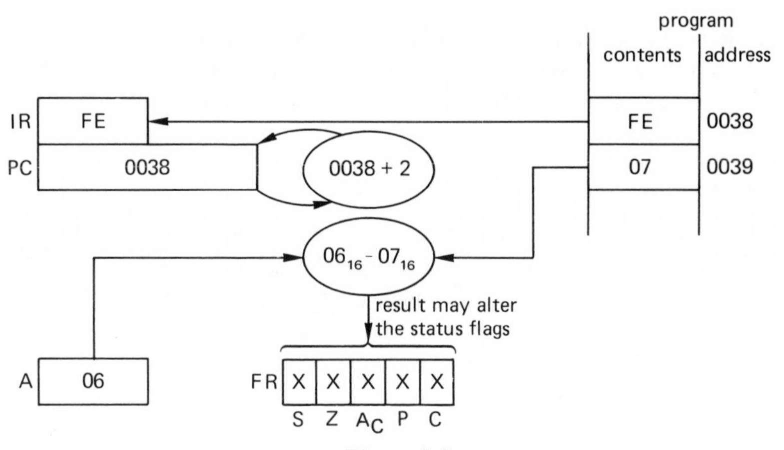

Figure 5.5

appropriate flags (the contents of the accumulator remain unchanged at 06_{16}). In the above case, the subtraction gives a non-zero result, the Z-flag (the zero flag) is reset.

Following the CPI instruction, the CPU executes a 3-byte **JZ** instruction (**J**ump to a specified address if the **Z**ero flag is set). If none of the keys has been pressed, the Z-flag is set to logic '1'. In this case, the JZ instruction causes control to be transferred to the address specified by the hexadecimal values in the second and third bytes of the instruction. That is, control is transferred to address 0030. However, if the CPI instruction results in the Z-flag being reset to logic '0' (that is, a key has been pressed), the JZ instruction is 'transparent' to the CPU and the next instruction obeyed is 003D, that is, step 2 of the program is executed, and the key which is pressed is identified.

Step 2: check which key has been pressed

This section is more complex than step 1, and only a brief description is given here. Referring to figure 5.3, the keyboard is scanned by driving **one** of the column wires low at a time (the other two remaining high), and then checking which row wire has a logic '0' on it. The logic '0' identifies which key has been pressed.

Keyboard scanning techniques often incorporate **two-key rollover** protection, which deals with the case where a second key is depressed before the first key is completely released. This form of protection ensures that keys are 'read' in the correct sequence.

5.4 Light-emitting Diode (LED) Displays

In this section we deal with the more popular forms of LED display which are

(1) annunciator lamps (single lamps)
(2) 7-segment displays

A LED is a semiconductor diode which generates a visible radiation when it is forward biased. The p.d. across a LED when it is forward biased is typically 1.6–2.5 V, and its current to give visible radiation is in the range 10–30 mA.

Single lamp display

The basis of a simple LED driver circuit is shown in figure 5.6; each LED is driven by an inverting buffer to provide sufficient current to illuminate the LED. A suitable inverting TTL IC is the 7416 which contains six inverting buffers, each of which can 'sink' or absorb a current of up to 40 mA at its output terminal; for this reason only six LEDs (L_0 - L_5) are connected to the DO terminals of the output port in the figure.

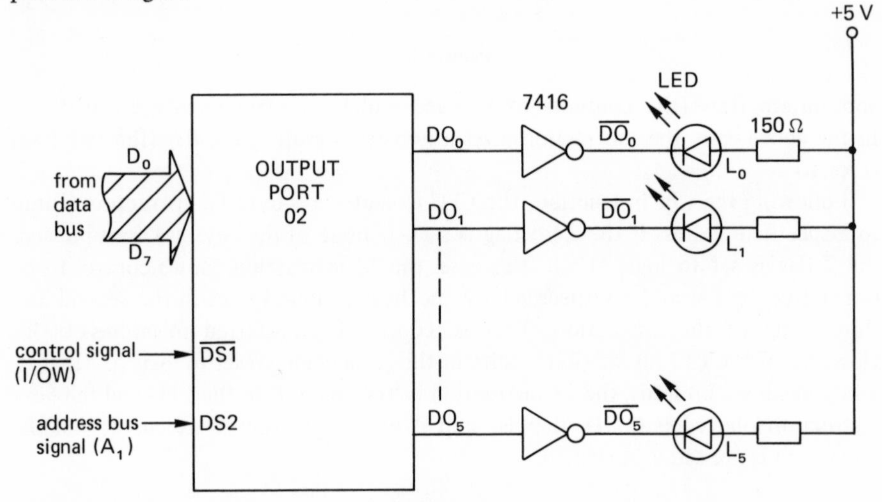

Figure 5.6

When the output port is selected by appropriate signals applied to the $\overline{DS1}$ and DS2 lines, the CPU data bus lines are connected to the DO lines of the output port. For the moment we will consider the operation of lamp L_0. When a logic '1' signal appears on output line DO_0, the inverting buffer provides an output logic '0'; this causes LED L_0 to be forward biased and it is therefore illuminated. When the signal on line $DO_0 = 0$, the output from the inverting buffer is logic '1'; in this case the p.d. across L_0 is low, so that L_0 is extinguished.

The application of a suitable binary word to the output port controls the operating condition of the lamps. The following sequence of instructions

```
3E    MVI 13₁₆
13
D3    OUT 02
02
```

causes the CPU to output the word 13_{16} (00010011_2) to the DO lines of output port 02. In turn, this causes LEDs L_0, L_1 and L_4 to be illuminated while L_2, L_3 and L_5 are extinguished.

The circuit in figures 5.6 can be modified so that a logic '0' turns a LED on, and a logic '1' extinguishes it, simply by replacing the 7416 IC by a 7417 TTL hex open-collector non-inverting buffer.

Seven-segment displays

A popular form of 7-segment display is shown in figure 5.7a. A LED is embedded at the bottom of a 'light pipe', which is a cavity filled with a transparent material. When the LED is energised, it results in a segment on the surface of the display being illuminated, each diode (corresponding to segments a to g, respectively) being energised independently. There are two alternative forms of the display,

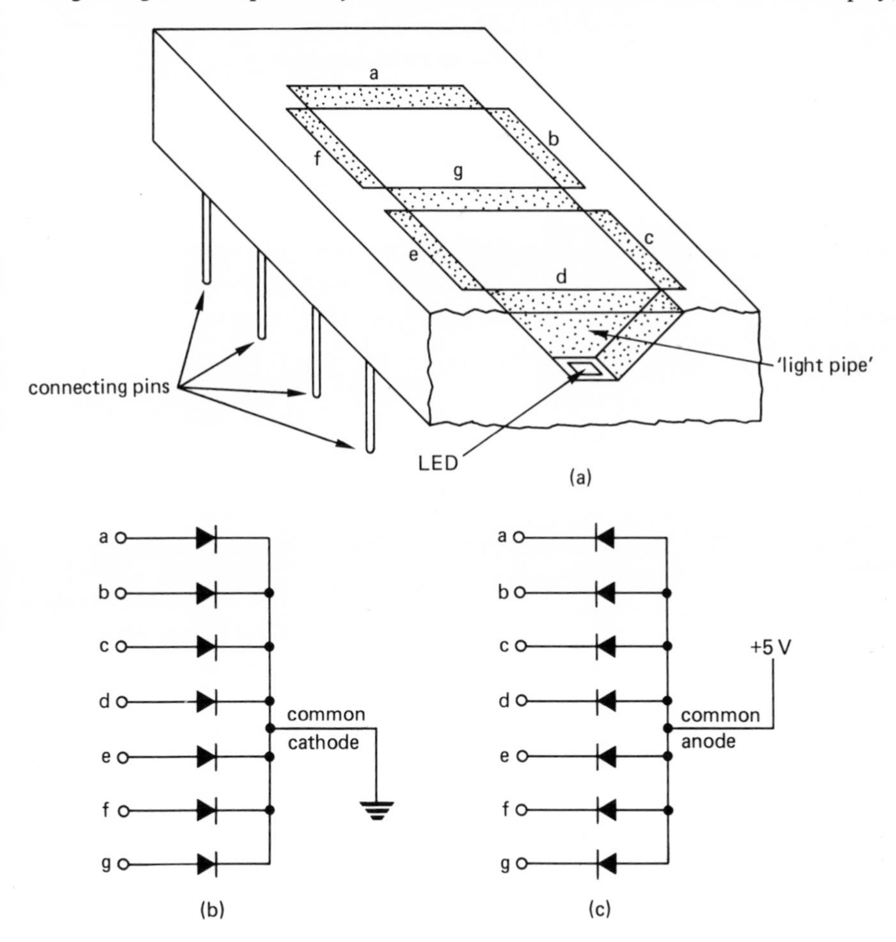

Figure 5.7

namely the common-cathode form (see figure 5.7b) and the common-anode form (figure 5.7c). In some cases an eighth LED is included to give a decimal point display. By illuminating various combinations of segments, the full range of hexadecimal characters can be displayed – see figure 5.8.

Figure 5.8

As with single LED lamps, the 7-segment display must have suitable drivers (buffers) together with current-limiting resistors (although in some cases these may be incorporated in the 7-segment display). In a common-cathode display, the 'common' pin is connected to the zero volts line, and each LED is illuminated by raising its input line to logic '1'. In a common-anode display, the common pin is connected to a positive supply rail, and each LED is illuminated by applying a logic '0' to its input line.

A block diagram showing a method of driving a common-cathode display is given in figure 5.9. Table 5.2 gives a range of binary words necessary to illuminate the hexadecimal characters in figure 5.8. Thus if the hex word 79 is output to a common-cathode display, the character E is illuminated. The reader will note that the data line DO_7 is not required in this display, and therefore bit 7 of the binary word can either be logic '0' or logic '1' without affecting the display. For this

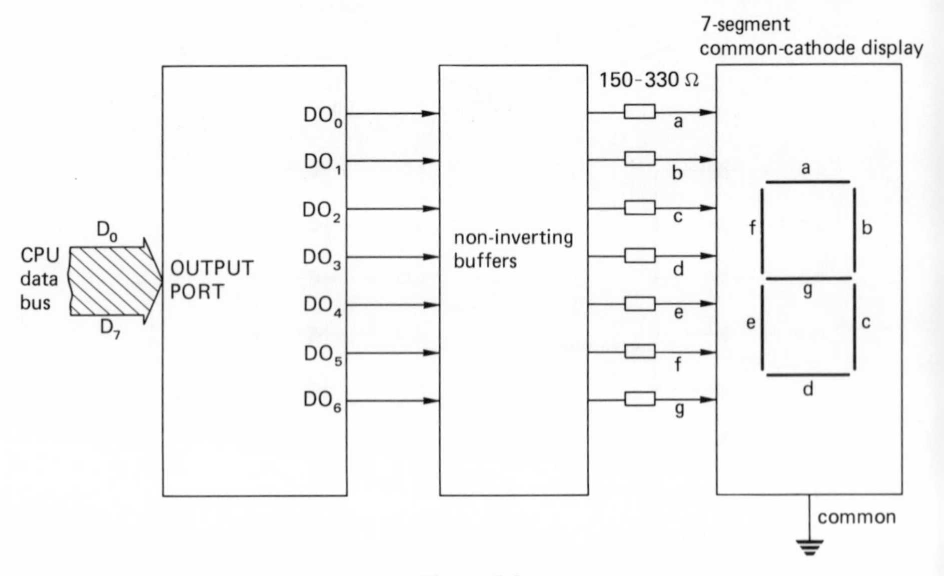

Figure 5.9

Table 5.2

Character displayed	Segments displayed	Binary word	Hexadecimal word
0	fedcba	00111111	3F
1	cb	00000110	06
2	gedba	01011011	5B
3	gdcba	01001111	4F
4	gfcb	01100110	66
5	gfdca	01101101	6D
6	gfedca	01111101	7D
7	cba	00000111	07
8	gfedcba	01111111	7F
9	gfcba	01100111	67
A	gfecba	01110111	77
b	gfedc	01111100	7C
C	feda	00111001	39
d	gedcb	01011110	5E
E	gfeda	01111001	79
F	gfea	01110001	71

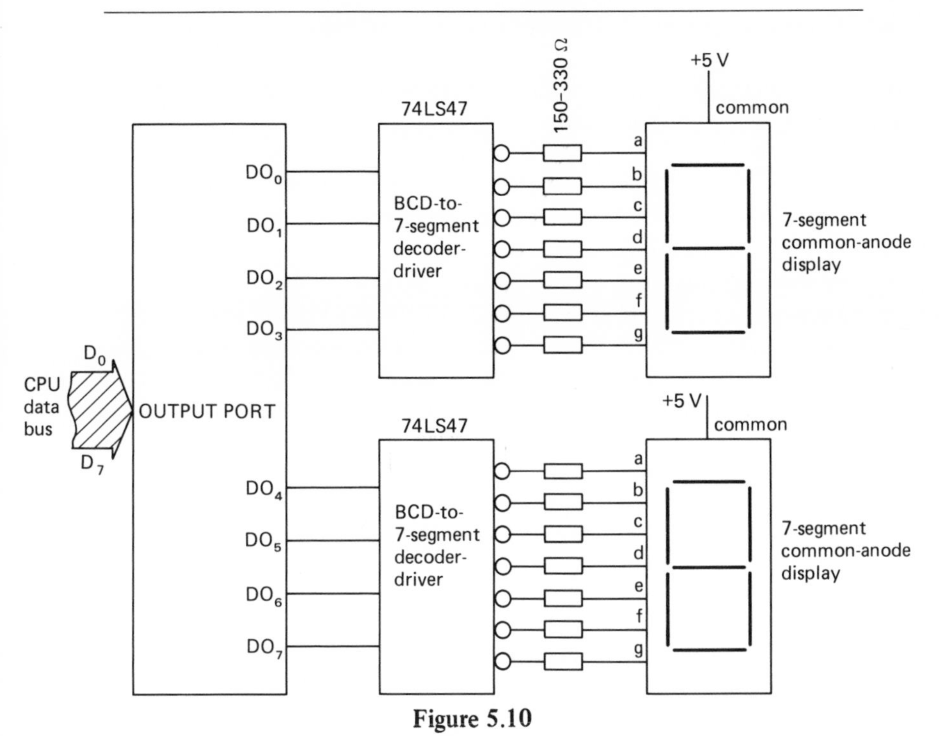

Figure 5.10

reason, if the hex word F9 (11111001_2) is output to the display, the character E would be displayed. Consequently, each hex character in the left-hand column of table 5.2 can be produced by two possible 8-bit words. However, if bit 7 is used to excite a decimal-point LED, then each combination in the right-hand column of table 5.2 uniquely defines the character in the left-hand column.

Figure 5.10 shows one method of driving 7-segment common-anode displays, but in this case the output port handles two 7-segment displays. The upper 74LS47 low-power Schottky decoder-driver decodes the BCD equivalent value on lines DO_0 - DO_3 into the correct combinations to drive the seven LEDs in the upper display. The reader will note that the output from the 74LS47 is inverted, so that if the signals on lines DO_0 to DO_3, inclusive, are logic '0' then the signals applied to the input lines of the display are $a = b = c = d = e = f = 0$ and $g = 1$. Bearing in mind that in a common-anode display a logic '0' illuminates a segment, then this combination illuminates the character 0 (refer back to figure 5.9). The reader may find it an interesting exercise to repeat table 5.2, but for a common-anode display.

The four lines DO_4 - DO_7, inclusive, are used in a similar fashion to drive the lower 7-segment display in figure 5.10.

PROBLEMS

5.1 Discuss the principal features of a monitor program.

5.2 In a microcomputer development system, give a range of addresses likely to be given to the monitor program.

5.3 Discuss the problems associated with contact bounce so far as they affect keyboards. Describe methods of overcoming the effects of contact bounce.

5.4 If the accumulator of a CPU contains 86_{16}, what changes are likely to occur in the state of the status flags (S, Z, A_C, P, C) after a CPI 52_{16} instruction.

5.5 What binary word must be output to a 7-segment display to illuminate (a) a negative (-) sign and (b) the letter S using (i) a common-anode display, (ii) a common-cathode display?

5.6 Draw up a table similar to table 5.2 but for a 7-segment common-anode display.

6 *Memory Organisation*

6.1 The Need for ROM and RAM

Read-only memories and random-access memories were briefly introduced to the reader in chapter 1. At this point we discuss the relevance of the two types of memory.

All microcomputers need some form of ROM. In the case, for example, of a general-purpose microcomputer, the ROM contains the monitor program which allows the user to input data from the keyboard or from some other device, and allows the CPU to output data to display devices. The ROM often contains a number of subroutines which are available to the user, that is, they are **user accessible**; a popular monitor subroutine is one allowing the user to shift a block of instructions from one area of memory to another. In the case of a dedicated microcomputer, the operating program is stored in ROM; when the microcomputer is first switched on, the CPU commences executing the program stored in ROM without the intervention of the operator. Read-only memories are available in alternative forms, known respectively as:

(1) ROM – **mask programmed** by the manufacturer
(2) PROM – **user programmable** (or **field-programmable**) read-only memory
(3) **EPROM** – **erasable PROM**

The above elements are described in this chapter.

The amount of RAM needed by a microcomputer depends on its application. A microcomputer dedicated to a very small system may only need a very small amount of RAM (say $\frac{1}{4}$K bytes); a system used for developing programs (a **development system**) for other small systems may need about 1K bytes of RAM; a general-purpose microcomputer used by a number of operators to perform a wide range of programs may need several thousand bytes of RAM.

To simplify the memory organisation of microcomputers, the computer store is divided into sections known as **pages**. In many microcomputers, a convenient page size is 256 ($\frac{1}{4}$K) words (corresponding to the hex range 00 – FF.) If, for example, the total storage capacity of the memory of a microcomputer is 2048 words (2K), then it has 8 pages of $\frac{1}{4}$K words. The method of page addressing is described later in this chapter. Other families of microcomputers use page sizes of $\frac{1}{2}$K words, 1K words and 4K words.

It is important to appreciate at this stage that a program stored in ROM is **non-volatile**, that is, the program (or data) is not lost when the power is switched off. However, the majority of RAMs are **volatile memories**, that is the program and data stored in them is lost when the power supply is removed. Some microcomputers have circuits which can protect vital data in the event of a power supply failure, but this is not generally the case.

6.2 A Basic Read-only Memory

The operation of a ROM is best explained in terms of a diode matrix of the type in figure 6.1. The matrix has four rows (R_0 - R_3) and eight columns (C_0 - C_7), some of the column wires being omitted for simplicity. The four row wires are energised via a decoder from the CPU address bus lines A_0 and A_1; the truth table of the decoder is given in table 6.1. By energising the address bus lines A_0 and A_1 with the appropriate combination of 1's and 0's, a logic '1' can be applied to any selected row wire. Thus when $A_0 = 0$ AND $A_1 = 0$, row wire R_0 has a logic '1' on it, the remaining row wires having 0's on them. When $A_0 = 0$ AND $A_1 = 1$, row wire R_2 has a '1' on it, and so on.

Table 6.1

Inputs		Outputs			
A_1	A_0	R_0	R_1	R_2	R_3
0	0	1	0	0	0
0	1	0	1	0	0
1	0	0	0	1	0
1	1	0	0	0	1

When a row wire has a logic '1' applied to it, the diodes connected to that line are forward biased, resulting in a logic '1' appearing on the appropriate column wire. For example, when $A_1 = 0$ AND $A_0 = 0$, the decoder circuit applies a '1' to row wire R_0; this causes diodes D1, D2 and D3 to be forward biased. The other diodes in the matrix (D4 - D11 are illustrated) are reverse biased (the reader may like to verify this statement). The net result is that a logic '1' appears on column wires C_0, C_1 and C_3; column wires C_2 and C_7 have logic 0's on them since no diodes are connected between R_0 and the column wires. It is not possible to say here what logic levels are applied to column wires C_4 to C_6, since it is not known if diodes are connected between them and R_0.

The eight column wires (C_0 - C_7) are connected to the eight data bus wires of the CPU. When row wire R_0 has a logic '1' on it, the 8-bit word specified by the diodes connected to R_0 is applied to the CPU data bus. Thus R_0 corresponds to one location in the ROM.

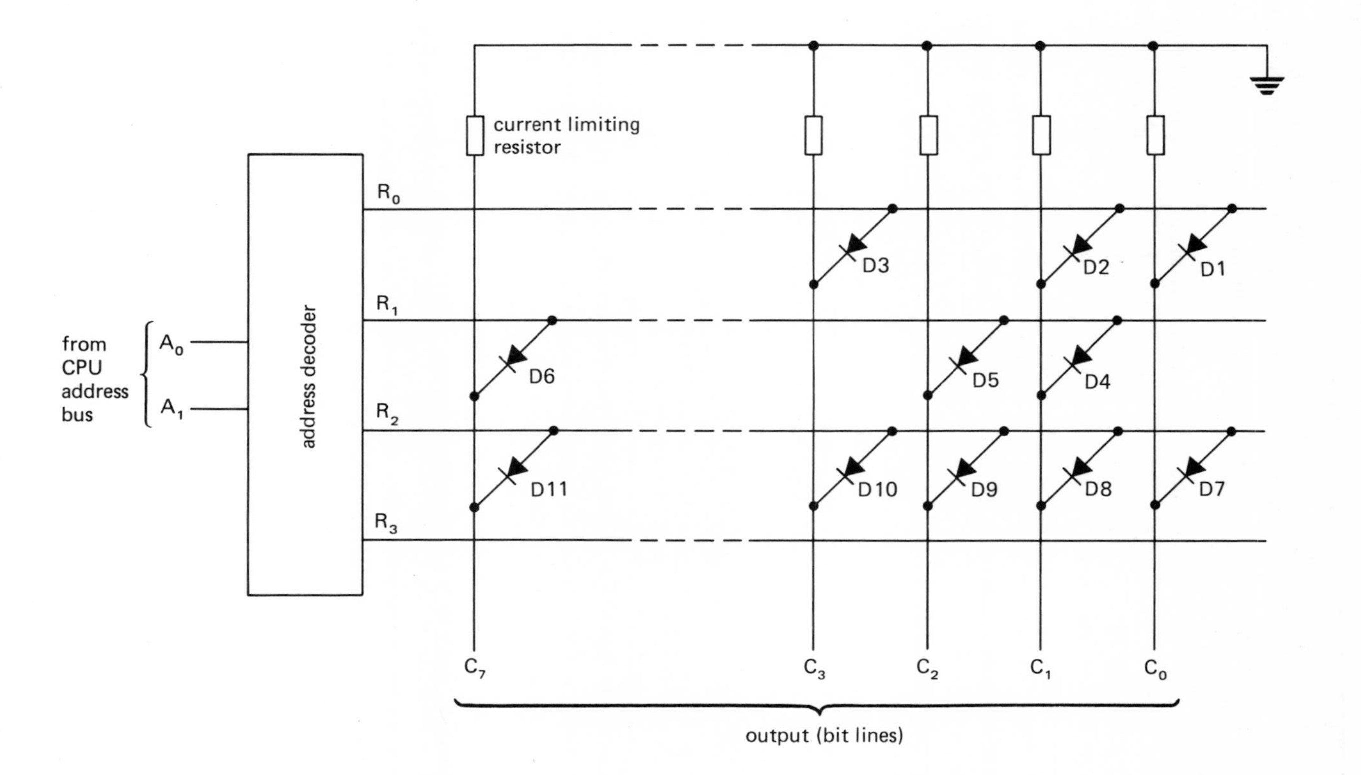
current limiting resistor
from CPU address bus
A_0
A_1
address decoder
R_0
R_1
R_2
R_3
D1
D2
D3
D4
D5
D6
D7
D8
D9
D10
D11
C_7
C_3
C_2
C_1
C_0
output (bit lines)

Figure 6.1

When $A_1 = 0$ AND $A_0 = 1$, row wire R_1 is selected. In this case, logic 1's appear on column wires C_1, C_2 and C_7, while 0's appear on wires C_0 and C_3.

If we assume that there is a diode at each column wire crossing with R_2, and there is none at each crossing with wire R_3, then the output word on the bit lines for $A_1 = 1$ AND $A_0 = 0$ is 11111111_2 (FF_{16}), and the output for $A_1 = 1$ AND $A_0 = 1$ is 00000000_2 (00_{16}).

Since two address lines (A_0 and A_1) are used in figure 6.1 to address the storage locations (the row wires), the ROM illustrated can store $2^2 = 4$ eight-bit words. If ten address lines ($A_0 - A_9$) are used (see also figure 5.1), then $2^{10} = 1024$ (1K) locations can be addressed. If each location stores one byte of information, then ten address lines allow the user to access each of the 1K 8-bit words.

Since each row wire addresses several column wires (eight in this case), each row wire is sometimes known as a **word address** wire. Each column wire is known as a **bit wire**.

The information stored in the ROM in figure 6.1 is established by the presence or absence of a diode at the junction of a row and a column wire. Clearly, the data (or instruction) can be altered simply by either connecting or removing a diode. However, this procedure is time consuming and the resulting ROM would be very bulky. The operation of a semiconductor ROM does not differ greatly in principle from that outlined above, but it differs in the method of achieving the result.

6.3 Mask Programmed Read-only Memory (ROM)

Using the monolithic integrated circuit technology described in chapter 3, the diode matrix in figure 6.1 can be formed at the manufacturing stage. Diodes would be produced at the junction of every row and column, with the exception that either the anode or the cathode of the diodes which are omitted in figure 6.1 are left open-circuited by the final production process.

However, it is more likely in a practical ROM that the diodes would be replaced by transistors. Both MOS and bipolar technology are in use.

6.4 Field-programmed ROM (PROM)

Figure 6.2 shows a simple **fusible link PROM** incorporating a diode matrix. The circuit is manufactured in monolithic IC form, each row wire being connected to each column wire by a diode and a polysilicon fuse. The customer receives the PROM with all the fuses intact so that, in the case shown, the output on each column wire is logic '1' when any of the word wires is energised by a '1'.

To program the memory the customer uses a **PROM programmer**, which is a portable piece of equipment that passes a high current pulse between selected word and bit lines in order to 'blow' appropriate fuses. In figure 6.2, if F_7 and F_0 are blown and F_1 is left intact, then a logic '1' applied to word wire R_0 causes logic 0's to appear on bit wires C_7 and C_0, and a '1' to appear on bit wire C_1.

The data used to program a PROM can be obtained from one of many sources

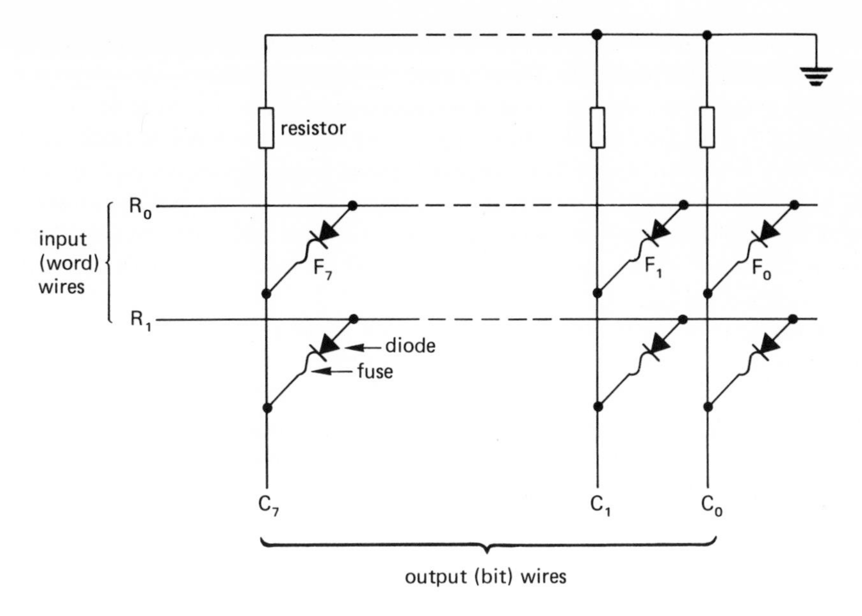

Figure 6.2

including a keyboard, or a paper tape, or a computer, or it can be copied from a 'master' memory which is already programmed. Certain PROM programmers contain a microprocessor as the controlling element.

Each fuse is blown by a pulse train of increasingly wider pulses, a current of 20–30 mA being typical of that required to blow the fuse. A temperature of about 1400 °C is reached at the point where the fuse blows, and at this temperature the polysilicon oxidises and forms an insulating material.

Other forms of fusible-link PROM exist, including those in which the diode is replaced by a transistor. Yet another form of PROM uses a reverse-biased diode in place of the fuse in figure 6.2; this is known as **shorted junction technology**. This type presents an open-circuit between the row and column wires, so that all column wires have logic 0's on them. To program this type of PROM, a high voltage pulse is applied between selected row and column wires in order to cause breakdown of the selected diode; this causes the selected diode to develop a permanent short-circuit. The shorted junction technology is not as popular as the fusible link construction, since it requires the use of a complex multiple-layer metalisation process.

6.5 Erasable PROM (EPROM)

EPROMS are electrically programmed by applying voltage pulses to the elements in the memory array and, if so desired, can be erased by some other means at a later date. This permits a memory chip to be reprogrammed to deal with modifications in the process being controlled.

A popular memory structure is the **floating-gate avalanche injection MOS (FAMOS)** shown in figure 6.3a. The structure is physically similar to the silicon-gate MOS transistor (see chapter 5 of *Semiconductor Devices* by Noel M. Morris, Macmillan Press, London). The silicon gate is surrounded by a silicon oxide (glass) layer, with the result that it is electrically isolated from its surroundings. Initially the gate region is charge-free (that is, no charge is stored in the gate) as shown in figure 6.3a. In this state, the source electrode is isolated from the drain electrode. This is illustrated in the circuit symbol (figure 6.3a) by the broken line between the source and the drain.

To turn the FAMOS ON, a voltage of 25 V or more is applied between the source and the drain. This causes avalanche breakdown of the *p–n* junction either at the source electrode or the drain electrode (depending on the polarity of the applied voltage); high-energy electrons in the avalanche region are injected into the floating gate when this occurs, where they are trapped since no leakage path exists from the gate. The amount of charge transferred to the gate is a function of the voltage applied to the device, and data is typically regarded as being transferred to a location in the EPROM only after about 100 voltage-application sequences; each scan causes a small amount of electrical charge to be 'written' into the location. The effect of the negative charge stored in the floating gate is to produce an **inversion layer** or *p*-type material in the *n*-type substrate. The inversion layer provides a conductive path between the source region and the drain region of the FAMOS. This conducting channel is indicated in the circuit symbol in figure 6.3b by the solid line linking the source and the drain.

The lifetime of the charge stored in the floating gate is a function not only of the operating temperature but also of the amount of electrical use of the EPROM. Tests carried out by one manufacturer suggest that at an operating temperature of 125 °C, the stored charge decays to about 70 per cent of its initial value in about 10 years. At lower temperature the decay rate is much slower.

The charge stored in the floating gate can be discharged by illuminating the FAMOS chip with high intensity ultraviolet light for about 20–30 min. To allow the user to do this, the EPROM chip is mounted behind a clear quartz window in the IC. The ultraviolet radiation results in the flow of photocurrent from the gate to the substrate, completely discharging the gate.

EPROM manufacturers recommend the use of a long erasure period with low intensity ultraviolet radiation rather than a shorter period with high intensity radiation; experience has shown that repeated erasure of an EPROM ultimately results in its being unable to store information.

6.6 Comparison of ROM, PROM and EPROM

The choice between a ROM, a PROM and an EPROM in a particular case is fairly clear cut.

An EPROM is the first choice where a program is being developed; the program can be modified easily by the user. PROMs and ROMs provide a more

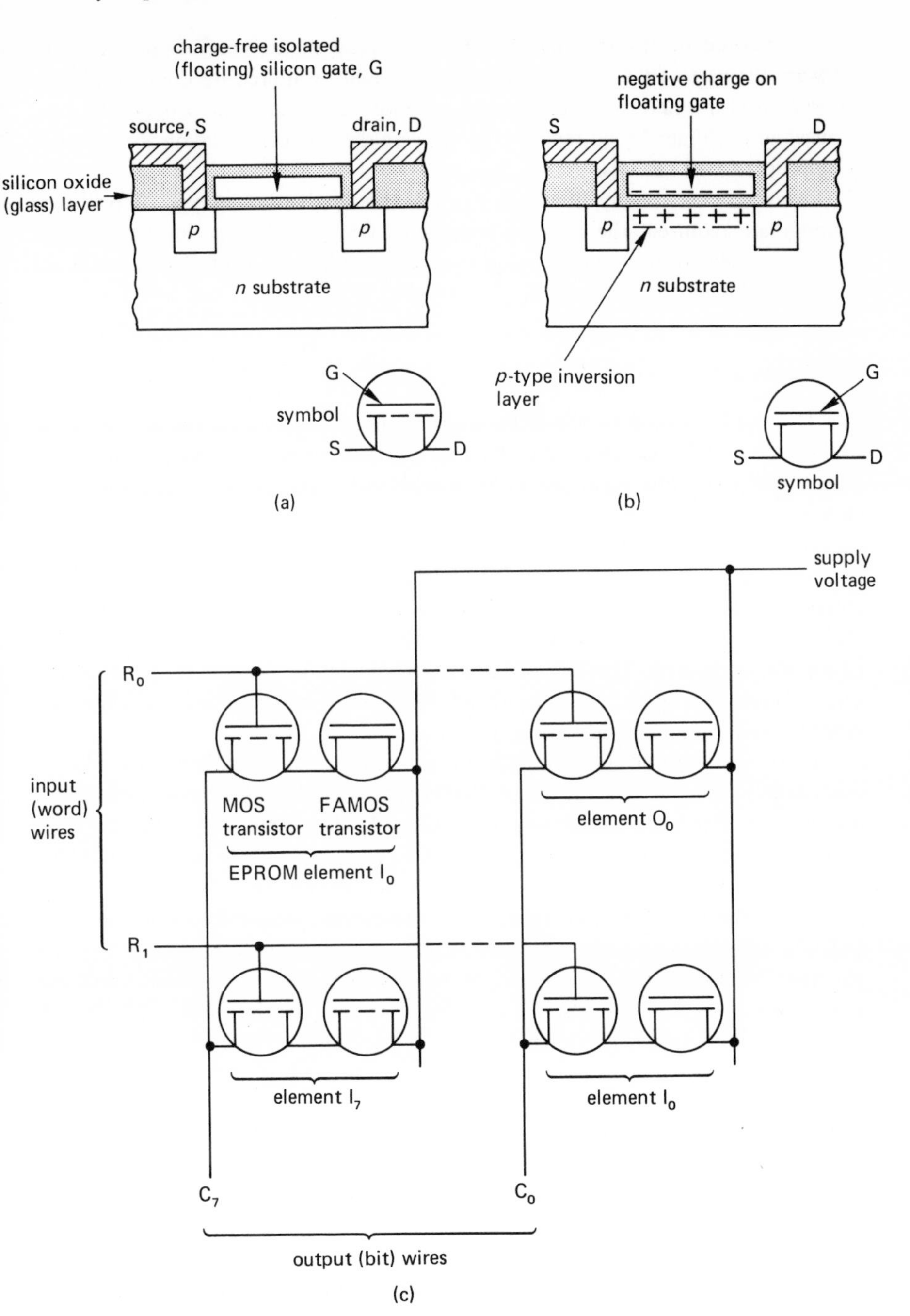
charge-free isolated
(floating) silicon gate, G
source, S
drain, D
silicon oxide
(glass) layer
p
p
n substrate
G
symbol
S
D
(a)
negative charge on
floating gate
S
D
p
p
n substrate
p-type inversion
layer
G
S
D
symbol
(b)
supply
voltage
R0
input
(word)
wires
MOS
transistor
FAMOS
transistor
EPROM element I0
element O0
R1
element I7
element I0
C7
C0
output (bit) wires
(c)

Figure 6.3

reliable method of storage than do EPROMs, since in the case of the latter the software (or part of it) can inadvertently be erased or altered by the user. PROMs are very useful where it is occasionally necessary to modify the program to meet changes in customers' requirements; in this case the user can maintain a small stock of unprogrammed PROMs.

However, where many hundreds or even thousands of one type of read-only memory are required, the mask-programmed ROM provides the lowest cost. Set against this advantage is the fact that the user may have to wait several weeks for the ROMs to be manufactured.

6.7 Electrical Connections to the ROM

The functions assigned to the ROM connection pins depend on the design of the ROM. The ROM described here illustrates popular practice, and has a storage capacity of 8192 bits organised in 1024 words of 8 bits. A block diagram of the chip is shown in figure 6.4.

Ten lines of the address bus (A_0-A_9) are used to address the 2^{10} = 1024 (1K) locations in the memory. Assuming that the ROM is organised on the row-and-column basis described earlier, the address lines specify row (word) wires from R_0 to R_{1023}, each row wire causing the eight bits of the selected word to be applied to the output buffers. The output buffers are enabled or selected by the two chip select signals $\overline{CS1}$ and $\overline{CS2}$. Data is therefore output to the data bus only when $\overline{CS1}$ AND CS2 are simultaneously low.

One of the chip select signals ($\overline{CS1}$) is energised from the address bus, and the other (CS2) from the control bus. A ROM of this kind was illustrated in the microprocessor in figure 5.1. Referring to figure 5.1, the reader will note that since address bus lines A_0-A_9, inclusive, are used to address locations within ROM, the remaining address bus lines lines (A_{10}-A_{15}) can be used for chip addressing. In figure 5.1, the $\overline{CS1}$ line is energised from the output of an OR-gate whose input lines are connected to address bus line A_{10} to A_{15}. Bearing in mind that the $\overline{CS1}$ pin must be driven low for the output buffers to be enabled, it follows that the address bus lines A_{10}-A_{15} must be low simultaneously for the ROM to be enabled. Thus the ROM in figure 5.1 stores data in addresses ranging from 0000_{16} to $03FF_{16}$ as follows

A_{15}	A_{14}	A_{13}	A_{12}	A_{11}	A_{10}	A_9	A_8	A_7	A_6	A_5	A_4	A_3	A_2	A_1	A_0
0	0	0	0	0	0	X	X	X	X	X	X	X	X	X	X
0_{16}				0-3_{16}				0-F_{16}				0-F_{16}			

where X = '0' or '1'. The 'start' address of the ROM is 0000_{16}, corresponding to row R_0 of the memory matrix; the 'finish' address of the ROM is $03FF_{16}$, corresponding to row R_{1023}.

Pin $\overline{CS2}$ is energised from the $\overline{MR}$ line of the control bus. The reader will recall that this line is forced low during the memory read cycle. Provided that the chip is

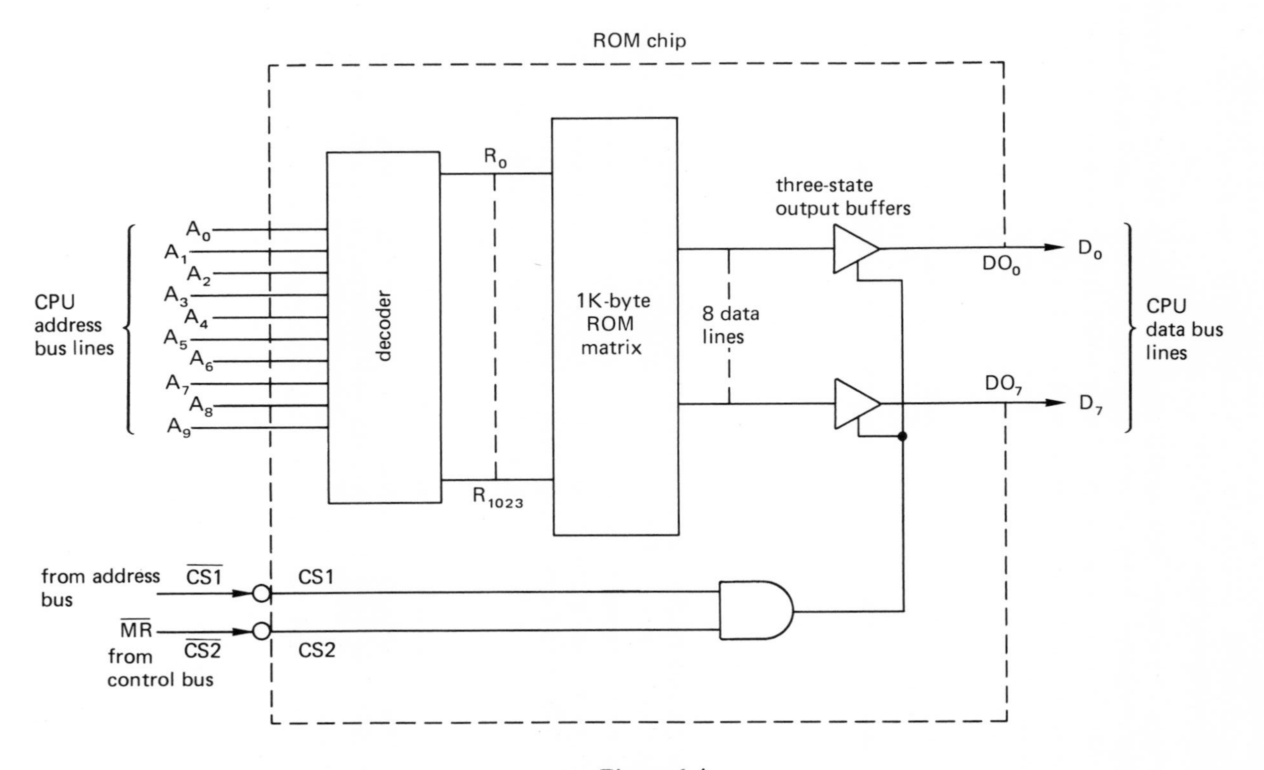

ROM chip
CPU
address
bus lines
A_0
A_1
A_2
A_3
A_4
A_5
A_6
A_7
A_8
A_9
decoder
R_0
R_{1023}
1K-byte
ROM
matrix
8 data
lines
three-state
output buffers
DO_0
DO_7
D_0
D_7
CPU
data bus
lines
from address
bus
$\overline{CS1}$
CS1
$\overline{MR}$
from
control bus
$\overline{CS2}$
CS2

Figure 6.4

addressed in the range 0000_{16} to $03FF_{16}$, then data will be output from the ROM during a memory read cycle.

6.8 Random Access Memories (RAMs)

In this type of memory, data can be either read from or written into a location addressed at random; RAMs are often described therefore as **read–write memories.**

A RAM is described as a **static memory** when data is stored in a conventional flip-flop, or is described as a **dynamic memory** when the data is stored in the form of an electrical charge on a capacitor. Both dynamic and static memories can be constructed using either bipolar or unipolar (FET) technolgoy.

6.9 Static RAMs

A simplified form of word-organised static RAM is shown in figure 6.5. The basic unit or memory cell is the S–R flip-flop containing the cross-connected transistors TR1 and TR2. In normal use the row (word) select wires are held at a low potential, so that the current in the transistor which is ON flows to the row wire.

To **read the stored data**, the appropriate row wire is raised to logic '1'. This causes the current in the transistor which is ON to flow to the appropriate column wire. In the case of the cell in the top left-hand corner of figure 6.5, the current flows to column wire C_0 if TR1 is ON, or it flows to wire $\overline{C_0}$ if TR2 is ON; this happens simultaneously to all the cells connected to wire R_0 when it is raised to logic '1'. In the case of the cell in the top left-hand corner of figure 6.5, it stores a logic '1' if TR1 is ON, and logic '0' if TR2 is ON.

When a location is 'read', the current in the bit lines flows into the input of the sense amplifier, the output signal from this being applied to a three-state buffer. The buffer gates are enabled when a logic '0' is applied to the $\overline{OD}$ line; the signal applied to this line is derived from the $\overline{MR}$ control bus line of the CPU (see also figure 5.1).

The reader will recall that when the CPU reads data from a memory, the $\overline{MR}$ control line is forced low; this simultaneously enables all the three-state output buffers of the memory, allowing stored data to be placed on the data bus. When the memory is not being read the $\overline{MR}$ control line is forced high, so disabling (or deselecting) the memory output.

To **write data** into a specific location, the CPU outputs the data on to the data bus, which is connected to the data input lines (DI_0 and DI_1 in figure 6.5) of the memory chip. During the memory write cycle, the $\overline{MW}$ control line is forced low which, together with the chip select signals (not shown) enables gates G1 and G2. Both of these gates have two outputs, one inverting (the upper output) and one non-inverting (the lower output). An input of logic '1' on line DI_0 causes column wire C_0 to be forced low and $\overline{C_0}$ to be forced high. At the same time, the row wire which is selected by the address signals is forced high; suppose that this is wire R_0. The logic '0' on wire C_0 forces transistor TR1 to the ON state, and the '1' on wire

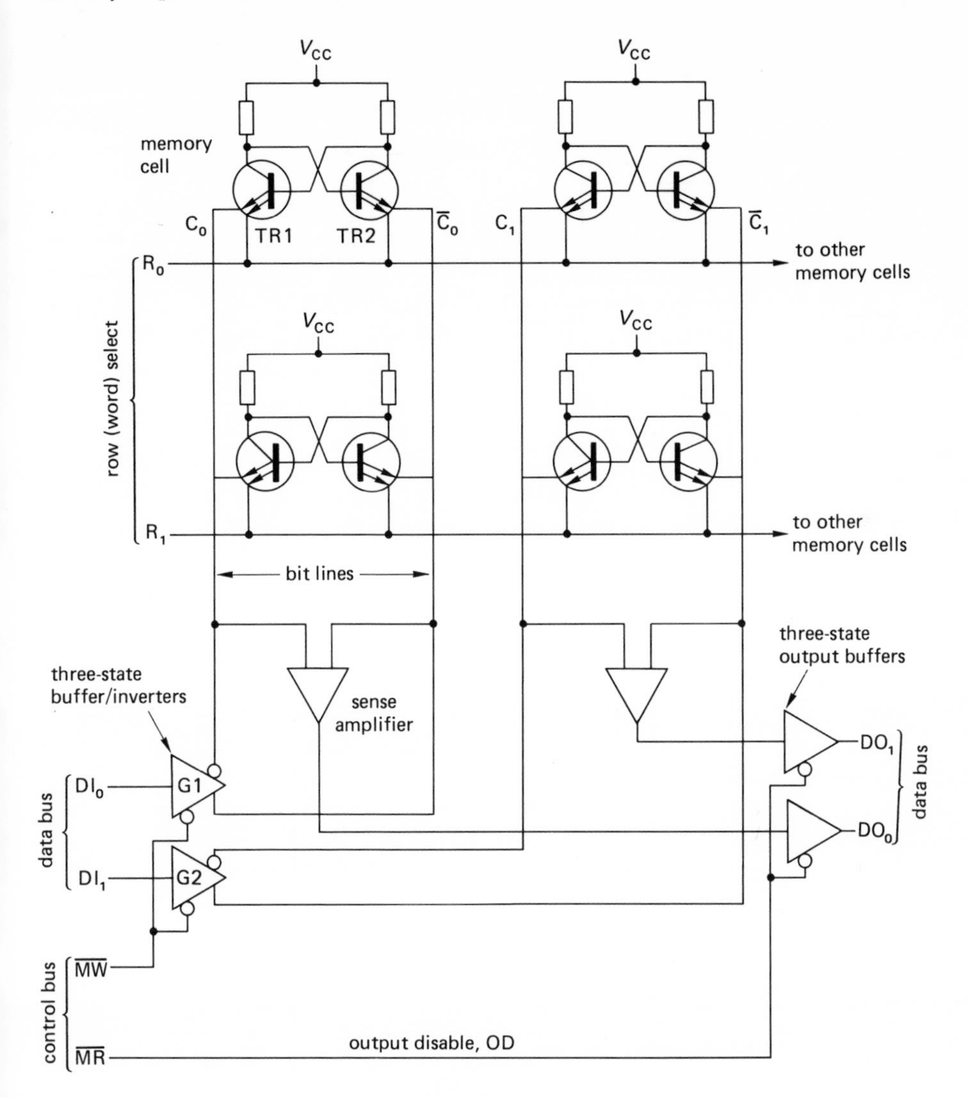

Figure 6.5

$\overline{C}_0$ forces TR2 OFF; thus the memory cell in the top left-hand corner of figure 6.5 stores a logic '1' after the above operation. After the completion of the 'write' sequence, row wire R_0 is forced back to logic '0' again, allowing TR1 to remain in its conducting state, so retaining the stored data. During the above operation the $\overline{MR}$ line remains high (see also figure 4.3), so that the output buffers are disabled.

The RAM shown in figure 6.5 has separate input and output lines. Many RAMs designed for microcomputer systems use a common set of connections for input and output of data. The basis of a 256 byte RAM chip with common I/O lines is

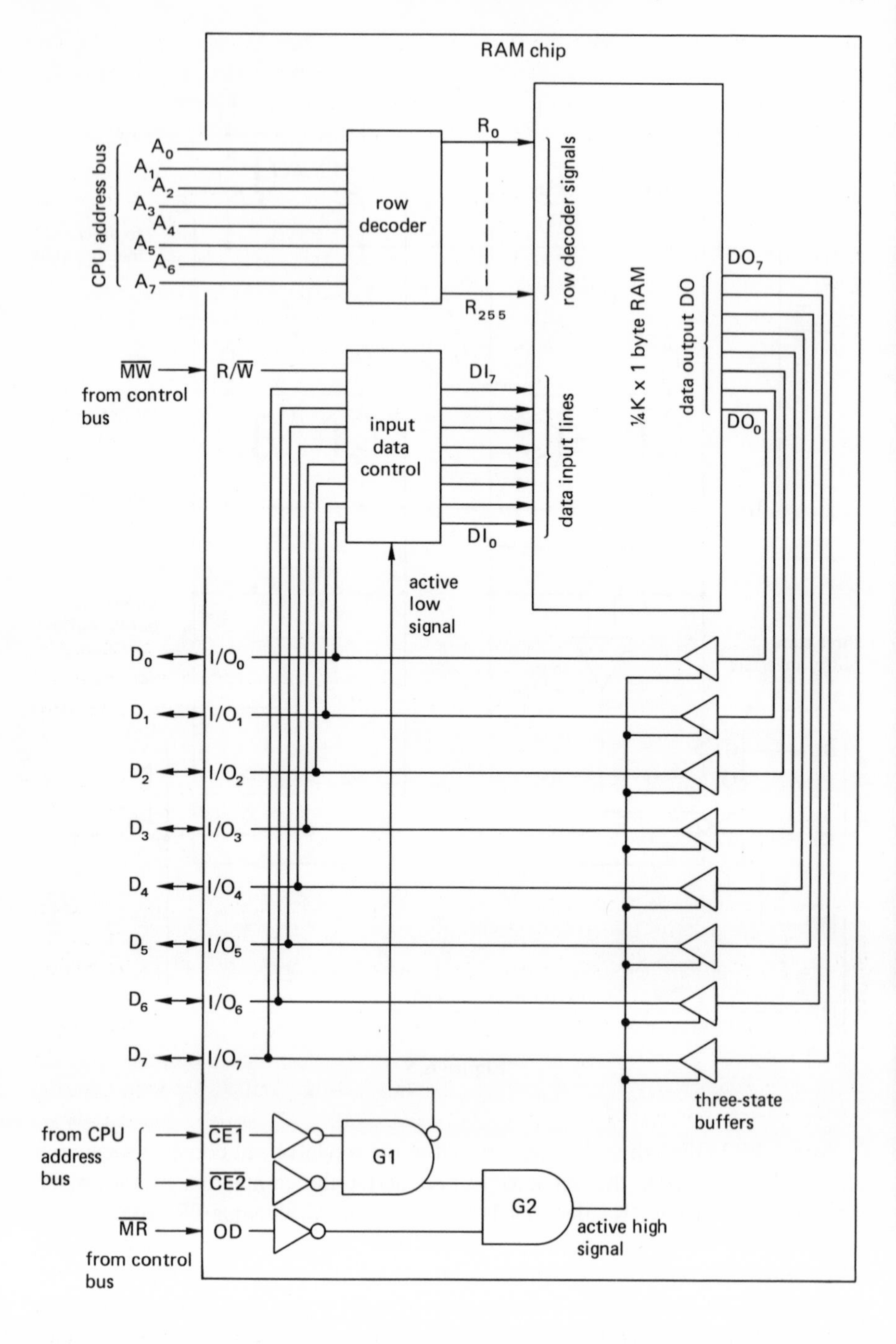

RAM chip
CPU address bus
A0
A1
A2
A3
A4
A5
A6
A7
row decoder
R0
R255
row decoder signals
¼K × 1 byte RAM
data output DO
DO7
DO0
MW
from control bus
R/W
input data control
DI7
DI0
data input lines
active low signal
D0 I/O0
D1 I/O1
D2 I/O2
D3 I/O3
D4 I/O4
D5 I/O5
D6 I/O6
D7 I/O7
three-state buffers
from CPU address bus
CE1
CE2
G1
G2
MR
OD
from control bus
active high signal

Figure 6.6

shown in figure 6.6 (see also the microcomputer block diagram in figure 5.1). The on-chip 'row' decoder in the figure decodes the signals on the CPU address bus lines A_0-A_7 to address one of the $2^8 = 256$ bytes of data in the RAM.

The block marked 'input data control' in figure 6.6 generally performs the function associated with gates G1 and G2 in figure 6.5. When the RAM pin assigned the symbol $R/\overline{W}$ is forced low, data can be written into the RAM (and is therefore energised by the $\overline{MW}$ control line of the CPU); when the $R/\overline{W}$ pin is held high, data cannot be written into the RAM. In some literature, this pin is assigned the symbol R/W; it is conventional to interpret this symbol as READ/NOT WRITE, that is, the letter W after the stroke is active low. Additionally, to permit data to be written into the RAM, the output from the inverting output of G1 must also be low. This occurs when $\overline{CE1}$ AND $\overline{CE2}$ are low simultaneously. The signals applied to the $\overline{CE1}$ AND $\overline{CE2}$ are obtained from the address bus in the manner shown in figure 5.1; taking figure 5.1 as an example of addressing, we note that

$$\overline{CE1} = A_8 + A_9$$

and

$$\overline{CE2} = \overline{A}_{10} + A_{11} + A_{12} + A_{13} + A_{14} + A_{15}$$

From the above logical relationships, the reader will note that $\overline{CE1}$ is taken low when A_8 and A_9 are low simultaneously, and that $\overline{CE2}$ is forced low when A_{10} is high and A_{11}-A_{15}, inclusive, are low. Hence the RAM chip in figure 5.1 is selected under the following conditions:

A_{15}	A_{14}	A_{13}	A_{12}	A_{11}	A_{10}	A_9	A_8	A_7	A_6	A_5	A_4	A_3	A_2	A_1	A_0
0	0	0	0	0	1	0	0	X	X	X	X	X	X	X	X
0_{16}				4_{16}				required for RAM addressing							

where 'X' could be a '0' or a '1'. That is, the RAM address is $04XX_{16}$ (the lower eight significant bits being used for addressing the 'word' lines or 'rows' in the RAM). For example, address 0400_{16} energises row R_0 in the RAM (see figure 6.6) and address $04FF_{16}$ energises row R_{255}.

When reading data from the memory, the $\overline{MW}$ line is held high, thereby inhibiting the operation of the input data control block. Also when reading data, the control bus line $\overline{MR}$ is forced low and, when taken in conjunction with the chip enable signal, causes the output of G2 to be high. This enables the three-state output buffers, allowing data to be output to the data bus from the address specified by the address bus information (A_0-A_7).

If the word length stored in the RAM is less than the word length of the microcomputer, it is possible to increase the apparent length of the RAM word by interconnecting several RAMs. An example of this is shown in figure 6.7. In this case, two 256 x 4-bit RAMs are used to store 256 bytes of data. Here, address lines A_0-A_7 are used to address the 256 locations, each storing a word length of 4 bits. The control and chip enable connections are generally as shown in figure 6.7, the internal organisation of the chips in figure 6.6 and figure 6.7 being identical (other than that the latter has a word length of 4 bits). The I/O lines of the upper RAM are connected to the four least significant lines of the data bus (D_0-D_3), while the I/O lines of the lower RAM are connected to data bus lines D_4-D_7. Since both chips are addressed simultaneously, data either is read from or is written into both chips at the same time.

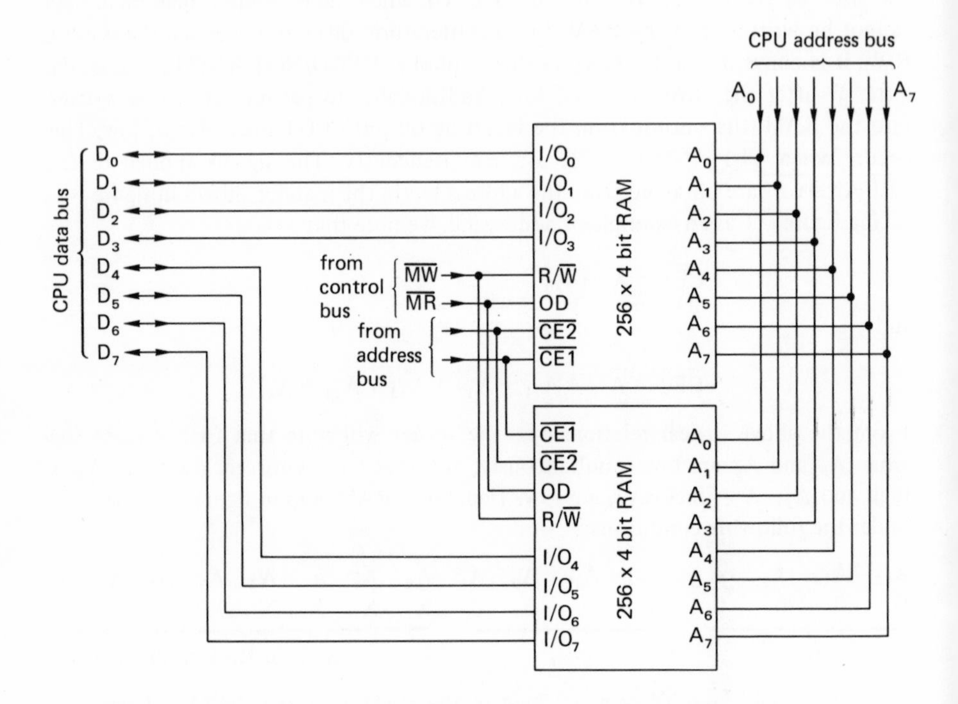

Figure 6.7

6.10 Dynamic RAMs

The basis of one form of cell used in a dynamic MOS RAM is shown in figure 6.8. The data is stored in the form of an electrical charge on the parasitic gate capacitors C_1 and C_2 of TR1 and TR2, respectively. With time, the charge on these capacitors leaks away, and it is necessary to provide circuits to 'refresh' the charge stored in them. For the sake of brevity the refresh circuits are not described here, but the basic method of writing data in the cell is outlined below.

In the following it is assumed that the cell stores a logic '1' when capacitor C_1 in figure 6.8 is discharged and C_2 is charged; when C_1 is discharged it maintains TR1 in the OFF or non-conducting state, and the charge on C_2 maintains TR2 in the ON or conducting state. The state of the cell is 'read' by energising row wire R_0; this action turns TR3 and TR4 ON simultaneously, connecting C_2 to column wire C_0

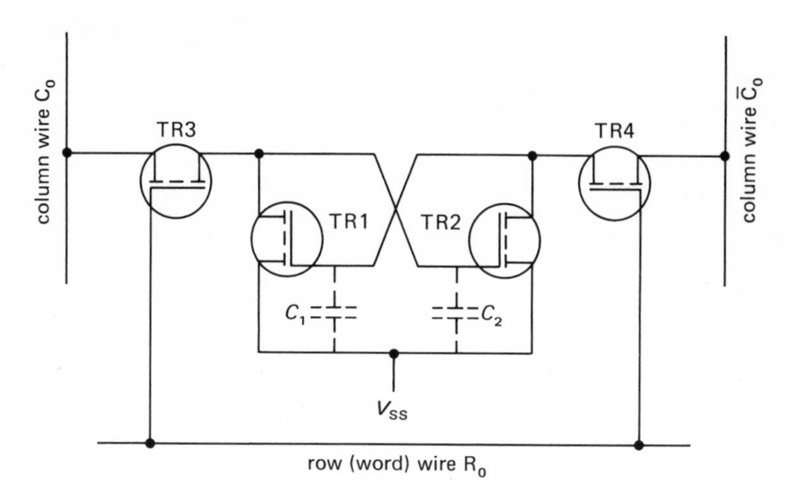

Figure 6.8

and C_1 to column wire $\overline{C_0}$. The differential conducting states of TR1 and TR2 indicate the logic value stored in the cell.

To 'write' data into the cell, TR3 and TR4 are turned ON once more and, at the same time, appropriate voltages are applied to column wire C_0 and $\overline{C_0}$; these voltages charge (or discharge) the parasitic gate capacitors C_1 and C_2 via TR4 and TR3, respectively.

6.11 Memory Expansion

The microprocessor discussed so far is assumed to have only one ROM and one RAM. In this section, methods of expanding the storage capability of the computer are discussed. Typical connections to 256 byte RAM and ROM chips are shown in figure 6.9a and b, respectively; what is of particular interest to us here is the signals applied to the CHIP SELECT pins.

The readers attention is directed towards the **page addressing** signal, which is applied to one (or more) of the CHIP SELECT pins of the ICs in figure 6.9; for an example of the way in which a ROM and a RAM are connected to a microprocessor system, the reader is referred to figure 5.1.

The memory page number referred to in figure 6.9 (and also elsewhere in the book) is given by the eight most significant bits (that is, the most significant byte) of the **absolute address** (this term is defined below). Thus if the address is $00XX_{16}$ (where X is any hex value in the range 0–F), then it is a 'page zero' address. If the address is $08XX_{16}$, then the address referred to is in page eight. The two least significant hexadecimal characters of the address (given as XX above) give the location within the page (the reader will recall that one page contains 256_{10} (or FF_{16}) locations). Thus address 0703_{16} refers to the fourth location in page seven.

A popular method of selecting any one of a large number of pages is by the use

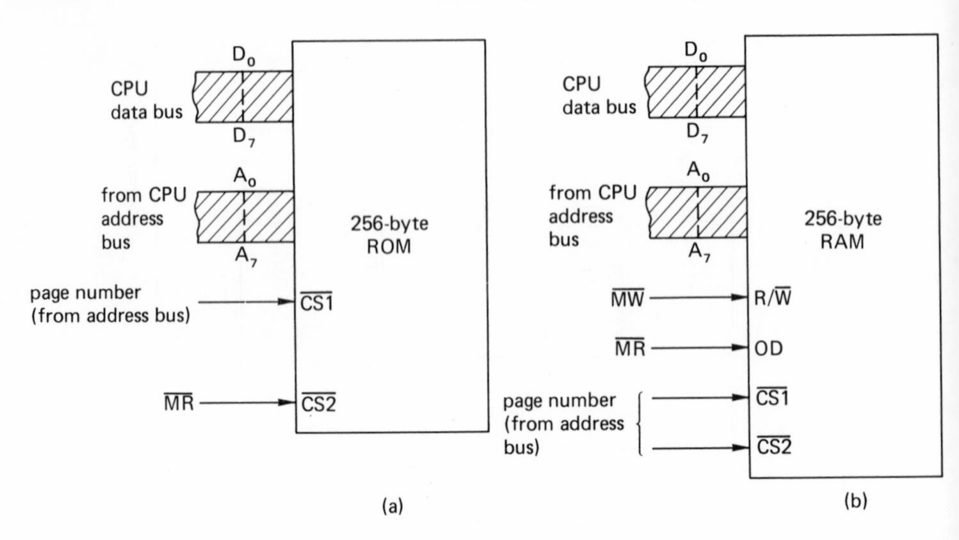

Figure 6.9

of decoders, a widely used decoder for this application being the 1-out-of-8 decoder (an example of this is the 8205 IC – a description of this type of IC is given in section 3.16).

The use of a 1-out-of-8 decoder is illustrated in figure 6.10. Assuming that each memory chip has 256 locations (that is, one page of storage capacity), then address lines A_0–A_7, inclusive, are needed to access the 256 'row' lines in each memory. The remaining address bus lines (A_8–A_{15}) can be used for the purpose of chip selection. The truth table for the 1-out-of-8 decoder in figure 6.10 is given in table 6.2. The truth table shows that when either E_1 or E_2 (or both) are high or when

Table 6.2 Truth table of a 1-out-of-8 decoder

Address			Enable			Outputs								
A_2	A_1	A_0	E_1	E_2	E_3	O_0	O_1	O_2	O_3	O_4	O_5	O_6	O_7	
0	0	0	0	0	1	0	1	1	1	1	1	1	1	
0	0	1	0	0	1	1	0	1	1	1	1	1	1	
0	1	0	0	0	1	1	1	0	1	1	1	1	1	
0	1	1	0	0	1	1	1	1	0	1	1	1	1	
1	0	0	0	0	1	1	1	1	1	0	1	1	1	
1	0	1	0	0	1	1	1	1	1	1	0	1	1	
1	1	0	0	0	1	1	1	1	1	1	1	0	1	
1	1	1	0	0	1	1	1	1	1	1	1	1	0	
X	X	X	1	X	X	1	1	1	1	1	1	1	1	chip disabled
X	X	X	X	1	X	1	1	1	1	1	1	1	1	chip disabled
X	X	X	X	X	0	1	1	1	1	1	1	1	1	chip disabled

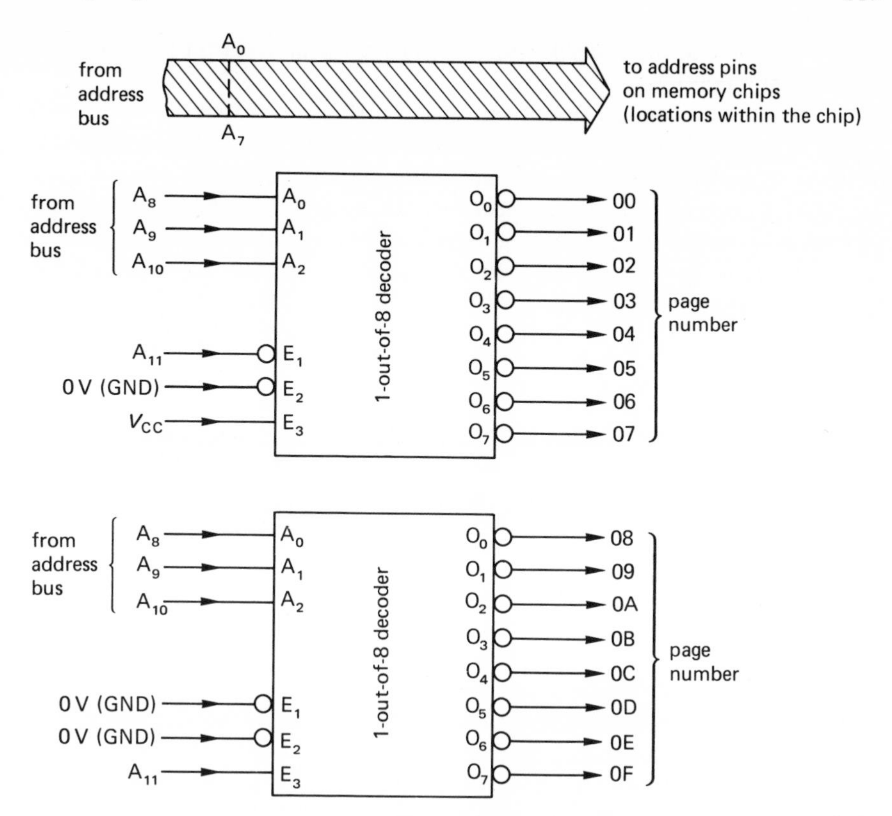

Figure 6.10

E_3 is low, then the chip is disabled, and all the outputs are logic '1'. When the chip is enabled (that is, $E_1 = 0$ AND $E_2 = 0$ AND $E_3 = 1$), a logic '0' appears at the output which is 'addressed' by the signals applied to the chip address lines A_0, A_1 and A_2. Thus if $A_2 = 0$, $A_1 = 1$, $A_0 = 1$ (corresponding to the number 011_2 or 3_{10}), a logic '0' appears at output O_3. In turn, this signal can be used to enable a memory chip (via a $\overline{CS}$ control pin) connected to the O_3 line of the decoder.

In the circuit in figure 6.10, the CPU address bus lines A_8, A_9 and A_{10} are connected to the chip address pins A_0, A_1 and A_2, respectively, on both decoders. As a result, when $A_{10} = 0$, $A_9 = 1$, $A_8 = 1$, an attempt is made to output a '0' from output O_3 of both decoders. However, as will be seen from the following, only one decoder at a time is enabled, and only the enabled IC can output a logic '0'.

In the case of both decoders, address line A_{11} is used to enable them; however, the upper decoder is enabled when $A_{11} = 0$ and the lower decoder is enabled when $A_{11} = 1$. The two remaining 'enabling' pins on each IC are connected to voltage levels which correctly enable the chip.

The addressing arrangement in figure 6.10 has the advantage of making page number 00_{16} one of the available addresses. This is useful where the starting address of the ROM is 0000_{16}.

The memory system associated with figure 6.10 can be expanded to 8K words by the addition of two decoders and another sixteen $\frac{1}{4}$K word memory chips (each of these chips being enabled by one of the outputs from the two additional decoders). The connections to the decoder inputs are as follows. Pins A_0, A_1 and A_2 on each decoder are connected to address bus lines A_8, A_9 and A_{10}, respectively; the connections to the 'enable' pins are as shown in table 6.3 in which logic '1' = V_{CC} and logic '0' = ground.

Table 6.3

Selector enable pin	Address line used to address the following page numbers			
	00 - 07	08 - 0F	10 - 17	18 - 1F
E_1	A_{11}	0	A_{11}	$\overline{A}_{11}$
E_2	A_{12}	A_{12}	0	0
E_3	1	A_{11}	A_{12}	A_{12}

The reader will note that address line $A_{12}-A_{15}$, inclusive, are not used in figure 6.10. However, if the memory chips can each store 1K words (such as the 8308 ROM), then address lines A_0-A_9 are needed to address the 1K addresses in each chip (note $2^{10} = 1024$). If an enabling arrangement similar to that in figure 6.10 is used, then 16 x 1K words = 16K words of memory can be addressed. In this case, address bus lines A_{10}, A_{11} and A_{12} are connected to the A_0, A_1 and A_2 pins, respectively, of the decoder chips; address line A_{13} would be used to energise the $\overline{E_1}$ input of the upper decoder in figure 6.10 and it would be connected to the E_3 line of the lower decoder.

6.12 Memory Maps

A memory map is simply a list of the addresses assigned to memory blocks in a computer system. To illustrate the formation of a memory map of a microcomputer, the readers should refer back to the microcomputer in figure 5.1. For the benefit of the reader, the relevant connections to the ROM and RAM chips are shown in figure 6.11a and b, respectively. The control bus signal connected to the $\overline{CS2}$ pin on the ROM and to the OD and R/$\overline{W}$ pins of the RAM have been discussed earlier in this chapter, and the reader should refer back to sections 6.7 and 6.9 for further information.

The locations of the ROM and RAM addresses on the memory map are decided by the way in which the address bus lines are connected to the individual chips. The reader will note that in the case of both ICs in figure 6.11, every address line

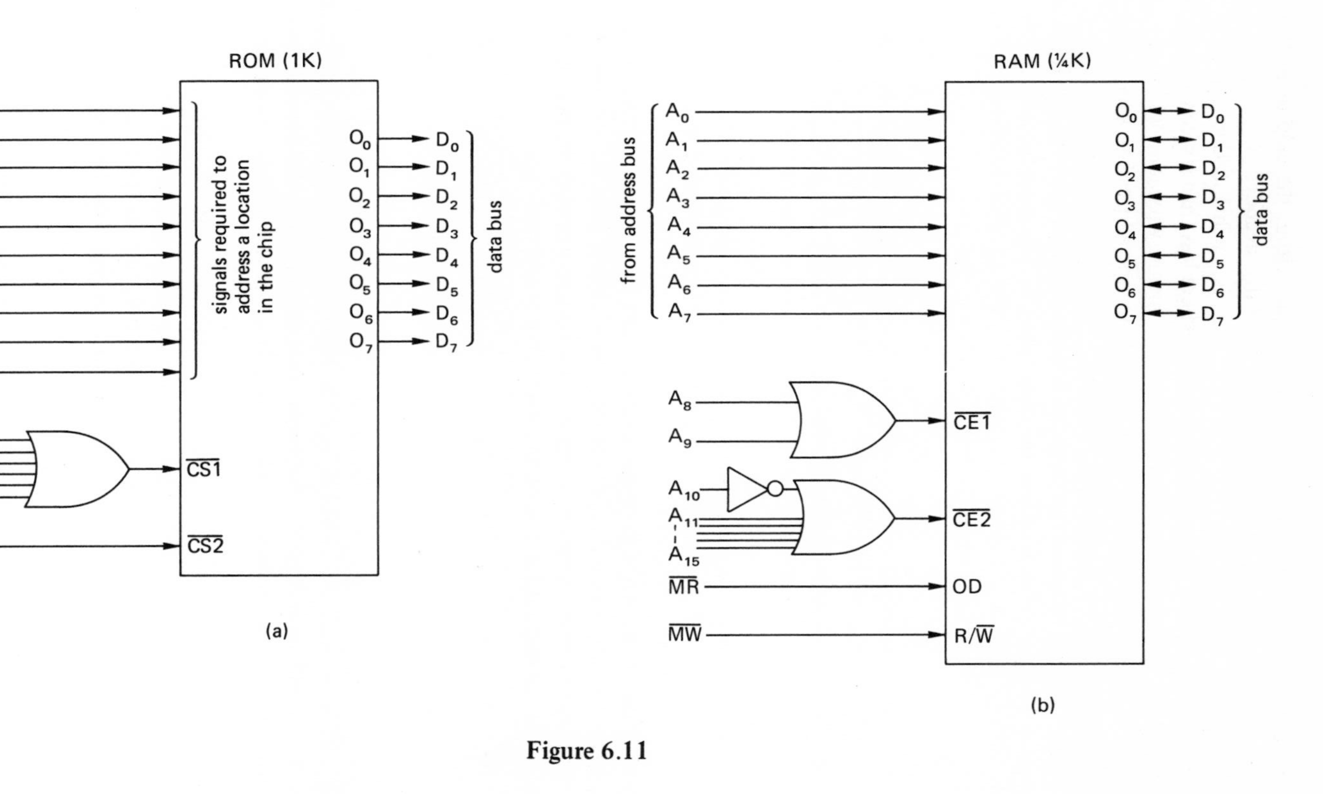
ROM (1K)
from address bus
A_0 A_1 A_2 A_3 A_4 A_5 A_6 A_7 A_8 A_9
signals required to address a location in the chip
O_0 O_1 O_2 O_3 O_4 O_5 O_6 O_7
D_0 D_1 D_2 D_3 D_4 D_5 D_6 D_7
data bus
A_{10} A_{15}
$\overline{CS1}$
$\overline{MR}$
$\overline{CS2}$
(a)
RAM (¼K)
from address bus
A_0 A_1 A_2 A_3 A_4 A_5 A_6 A_7
O_0 O_1 O_2 O_3 O_4 O_5 O_6 O_7
D_0 D_1 D_2 D_3 D_4 D_5 D_6 D_7
data bus
A_8 A_9
$\overline{CE1}$
A_{10} A_{11} A_{15}
$\overline{CE2}$
$\overline{MR}$
OD
$\overline{MW}$
$R/\overline{W}$
(b)

Figure 6.11

(A_0–A_{15}) is involved either in selecting an address within the chip or in the chip selection process, that is the address lines are **fully decoded.** Consequently the address of each memory location is unique, that is, each location can have one address only. As will be seen later, in the case where less than sixteen address lines are involved in the selection of the chip and the memory location within the chip, then each memory location may have more than one address; in such a case the chip addressing is said to be **partially decoded.**

Consider for the moment the range of addresses associated with the ROM chip in figure 6.11a. In order to select the chip, the $\overline{CS_1}$ pin must be low; for this to occur, address lines A_{10} through A_{15} must be logic '0' simultaneously. The logic levels on the address lines when the ROM chip is selected are

A_{15}	A_{14}	A_{13}	A_{12}	A_{11}	A_{10}	A_9	A_8	A_7	A_6	A_5	A_4	A_3	A_2	A_1	A_0
0	0	0	0	0	0	X	X	X	X	X	X	X	X	X	X
0_{16}				0–3_{16}				0–F_{16}				0–F_{16}			

where X = '0' or '1'. Thus to address the lowest (first) location in ROM, each address line must have a logic '0' on it, that is, the address is 0000_{16}. The highest address in ROM is accessed when address lines A_0–A_9, inclusive, have logic 1's on them; that is, its address is $03FF_{16}$. Hence the range of addresses associated with the ROM in figure 6.11a is 0000_{16} to $03FF_{16}$.

In the case of the RAM in figure 6.11b, address lines A_0 through A_7 are needed to address the locations within the RAM, while address lines A_8 through A_{15} are used in the chip enabling process. Applying a similar argument to that above, the states of the address bus lines when the RAM chip is addressed are

A_{15}	A_{14}	A_{13}	A_{12}	A_{11}	A_{10}	A_9	A_8	A_7	A_6	A_5	A_4	A_3	A_2	A_1	A_0
0	0	0	0	0	1	0	0	X	X	X	X	X	X	X	X
0_{16}				4_{16}				0–F_{16}				0–F_{16}			

The lowest (first) available location in the RAM is addressed when address lines A_0–A_7 all have logic 0's on them, and the highest available location is addressed when A_0–A_7 all have logic 1's on them. Additionally, address lines A_8, A_9 and A_{11}–A_{15} must be low while A_{10} must be high. That is, the RAM addresses range from 0400 to 04FF.

Since the microprocessor considered here uses only one ROM chip and one RAM chip, both with fully decoded addresses, then the range of addresses 0500 to FFFF are available for memory expansion. The memory map of the microcomputer in figure 5.1 is therefore as shown in figure 6.12; in this case the page size is 256 words. The same information is presented in figure 6.13 in a memory map using a 4K word page size.

An example of a memory system in which less than fifteen lines are used to select the chips is illustrated in figure 6.14. The ROM in figure 6.14a has a storage capacity of 512 words, and therefore requires nine address lines (A_0–A_8) to select addresses within the chip. In this case the designer has chosen to use only address

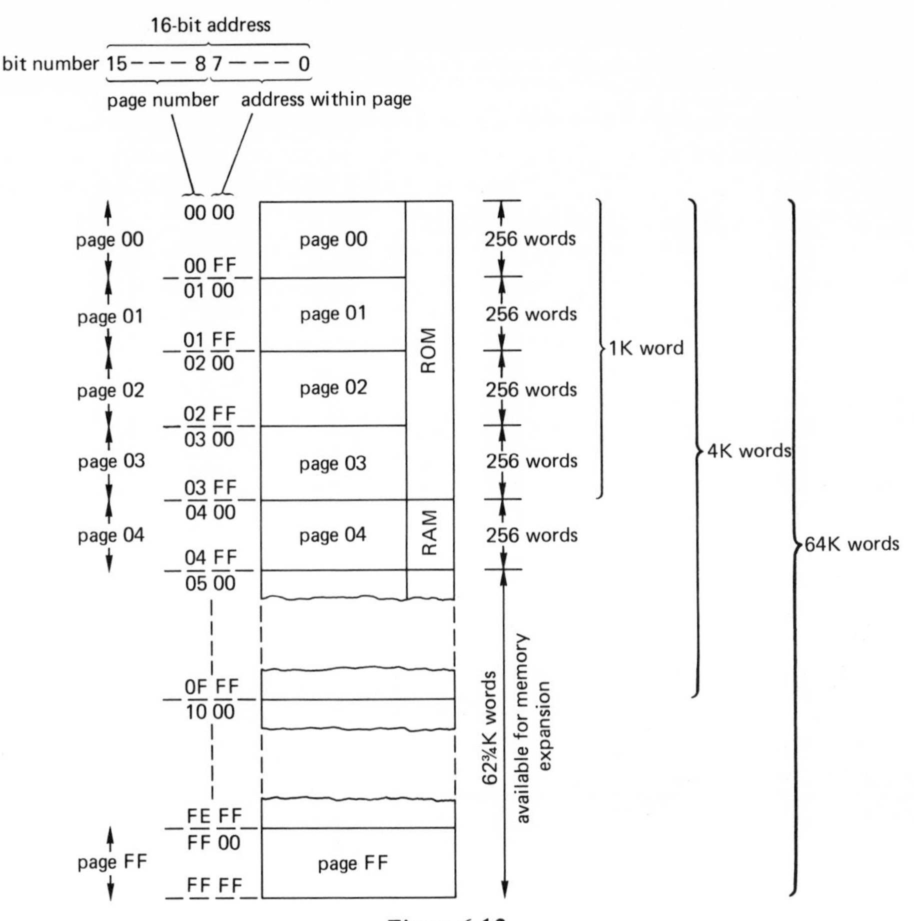

Figure 6.12

line A_9 and A_{10} to energise the $\overline{CS1}$ chip select pin. Since the two signals are OR-gated to the $\overline{CS1}$ pin, both of these address lines must be low in order to select the ROM. The logic signals on address lines A_{11} through A_{15} are immaterial since they are not connected to the chip. Thus the ROM is selected when the following conditions occur on the address lines

A_{15}	A_{14}	A_{13}	A_{12}	A_{11}	A_{10}	A_9	A_8	A_7	A_6	A_5	A_4	A_3	A_2	A_1	A_0
X	X	X	X	X	0	0	X	X	X	X	X	X	X	X	X
$0\text{-}F_{16}$				$0, 1, 8$ or 9_{16}				$0\text{-}F_{16}$				$0\text{-}F_{16}$			

(A_8–A_0: required for an address within the chip)

The reader will note that address lines $A_0 - A_8$ are used for in-chip memory addressing. Since address line A_{11} is not used, it may have either a logic '0' on it or a logic '1' on it without affecting the address selected within the chip; this comment also

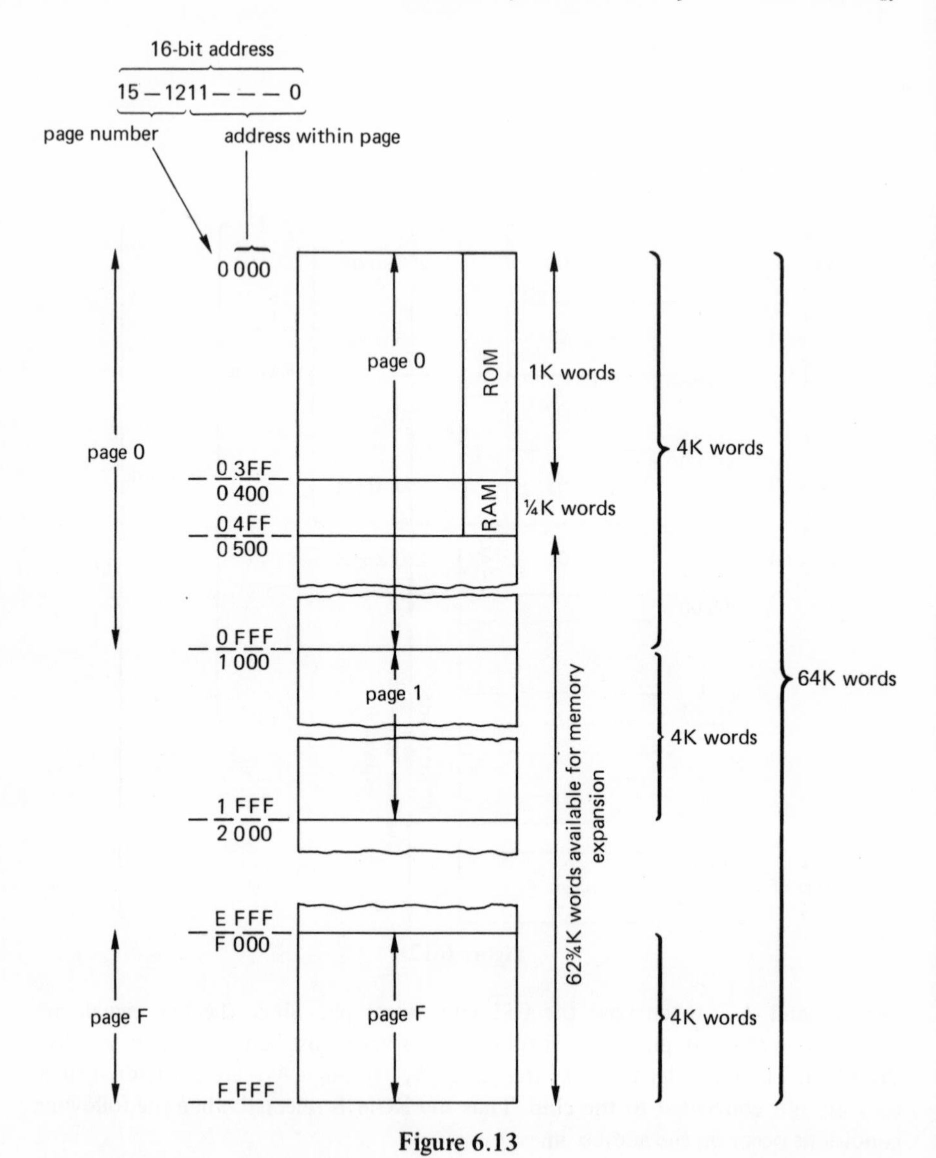

Figure 6.13

applies to address lines A_{12}–A_{15}. If address lines A_{11}–A_{15} each have logic 0's on them, the address of the lowest (first) location in ROM is 0000_{16}, and that of the highest location is 01FF, that is, there are $1FF_{16}$ (512_{10}) locations in ROM.

However, since address lines A_{11}–A_{15} can have 1's or 0's on them, the first location in ROM can have any of the hex addresses 0000, 0800, 1000, 1800, 2000, 2800, 3000, 3800, . . . F000, F800. That is, each location has thirty-two apparent addresses, any of which may be used to access the location.

The readers attention is now directed to the RAM in figure 6.14b. In this case

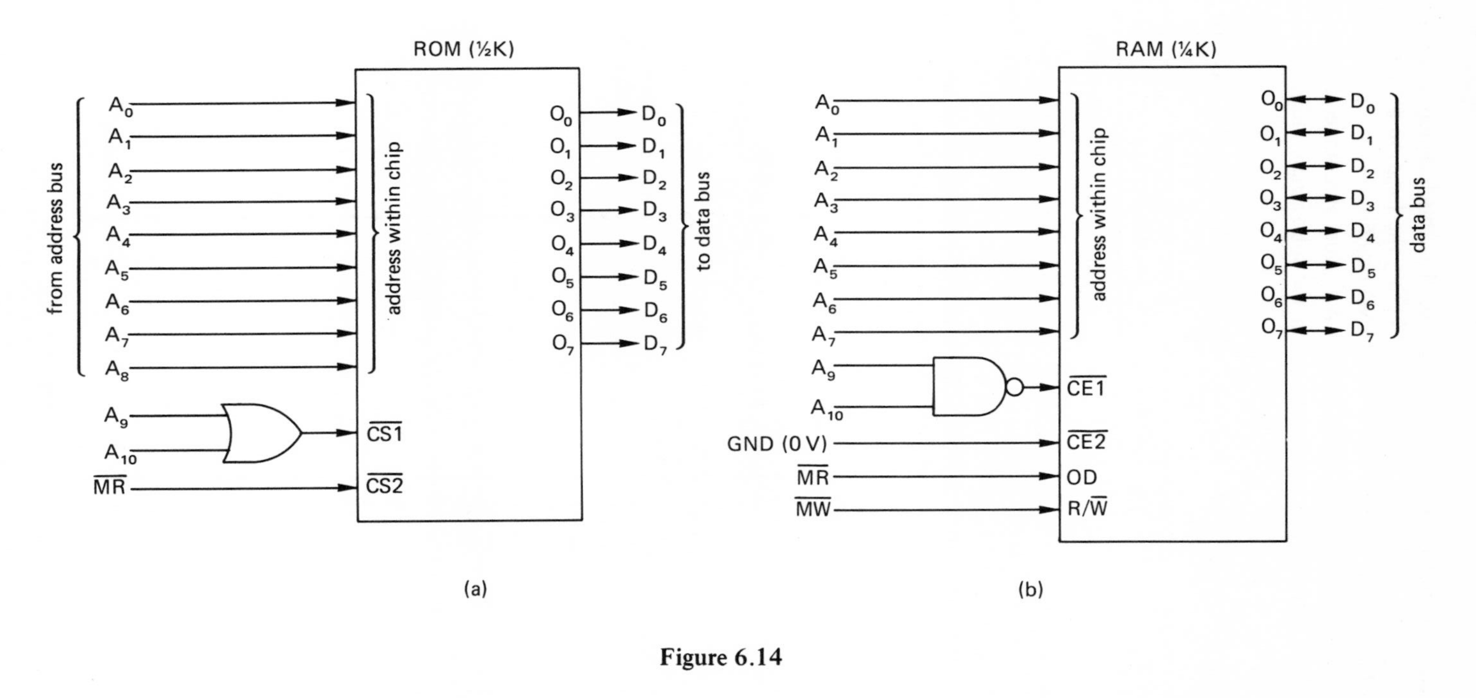
ROM (½K)
from address bus
address within chip
to data bus
A_0 A_1 A_2 A_3 A_4 A_5 A_6 A_7 A_8
A_9 A_{10} $\overline{MR}$
$\overline{CS1}$ $\overline{CS2}$
O_0 O_1 O_2 O_3 O_4 O_5 O_6 O_7
D_0 D_1 D_2 D_3 D_4 D_5 D_6 D_7
(a)
RAM (¼K)
address within chip
data bus
A_0 A_1 A_2 A_3 A_4 A_5 A_6 A_7
A_9 A_{10}
GND (0 V)
$\overline{MR}$ $\overline{MW}$
$\overline{CE1}$ $\overline{CE2}$ OD $R/\overline{W}$
O_0 O_1 O_2 O_3 O_4 O_5 O_6 O_7
D_0 D_1 D_2 D_3 D_4 D_5 D_6 D_7
(b)

Figure 6.14

the designer has not used address line A_8, so that this line may have either a logic '1' or a logic '0' applied to it without affecting the location selected in the chip. The signal applied to the $\overline{CE1}$ pin is the NAND function of the signal on the A_9 and A_{10} lines; to drive $\overline{CE1}$ low, both A_9 and A_{10} must simultaneously be high. Thus the RAM in figure 6.14b is selected when the following conditions occur on the address lines

A_{15}	A_{14}	A_{13}	A_{12}	A_{11}	A_{10}	A_9	A_8	A_7	A_6	A_5	A_4	A_3	A_2	A_1	A_0
X	X	X	X	X	1	1	X	X	X	X	X	X	X	X	X
$0-F_{16}$				6, 7, E or F_{16}				$0-F_{16}$				$0-F_{16}$			

required for an address within the chip (A_7 through A_0)

The first (lowest) location in the RAM chip (addressed when A_0 through A_7 are at logic '0') can be accessed at any one of the following hex addresses: 0600, 0700, 0E00, 0F00, 1600, 1700, 1E00, 1F00, etc.

The memory map for the ROM/RAM arrangement in figure 6.14 for the first 4K locations is as shown in figure 6.15; this map is repeated in each 4K word group up

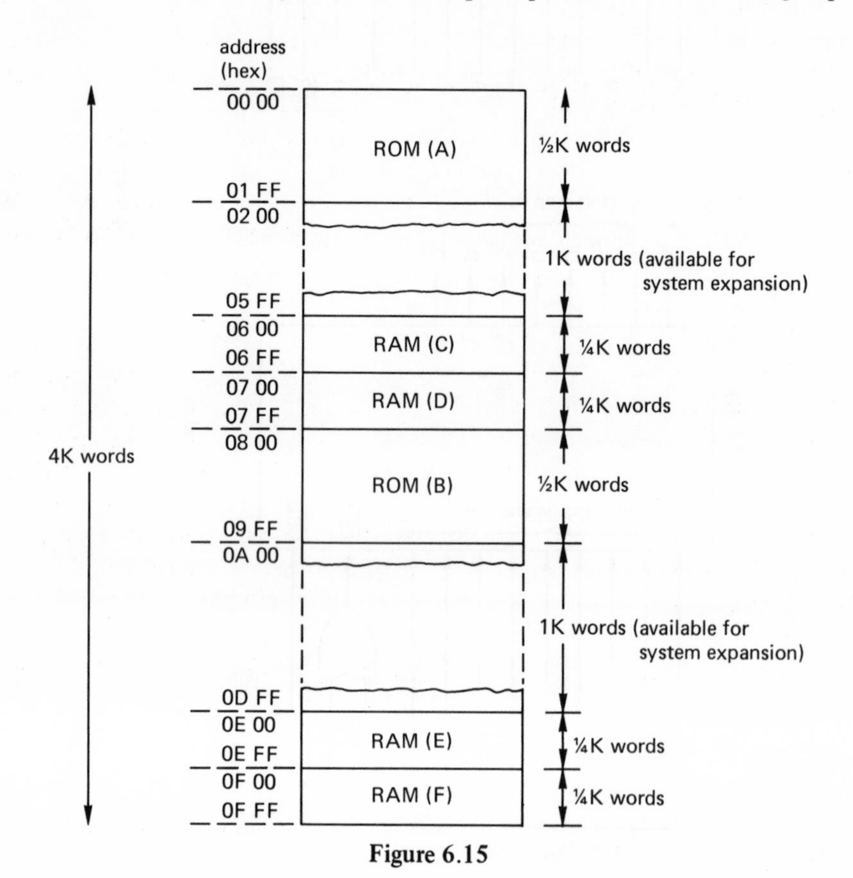

Figure 6.15

to the maximum of 64K words. The map shows that in each 4K word group, the data in each ROM location can be found at any one of two possible addresses, once within the ROM(A) block and once within the ROM(B) block, that is, the contents of address 0020 are also to be found at address 0820. The contents of each RAM location are repeated in each of the four blocks RAM(C), RAM(D), RAM(E) and RAM(F), that is, the contents of address 0631 are also to be found at 0731, 0E31 and 0F31.

The reader will appreciate that while incomplete decoding of the chip select signal is inefficient in terms of the use of large areas of the memory map (the decoding in figure 6.15 uses 2K words of each 4K word group), it leads to very simple decoding circuitry. Moreover, 2K words in each 4K word group remain available for memory expansion.

6.13 Relationship between the Size of RAM and ROM

The relative size of RAM and ROM is a function of the projected use of the microcomputer. However, there is a lower limit to the size of either type of memory below which they are impractically small. In general, the smallest useful size of RAM is about 256 words; the smallest size of ROM into which a useful monitor program can be fitted is about $\frac{1}{2}$K words (the reader may, of course, be aware of exceptions to these comments).

An international CPU manufacturer has carried out an investigation into the relative sizes of ROM and RAM requirements, and has found that many systems need a ROM storage capacity which is four to eight times larger than the RAM capacity. This ratio is subject to the lower limitations of size mentioned above, and to the application of the microcomputer.

6.14 Direct Memory Access (DMA)

So far in the book, data transfers between a peripheral device and a storage location have been handled by the CPU. In the case of a transfer from, say, a magnetic disc store to a RAM, each byte of data is input via an input port to the accumulator, and is then transferred to a specified storage location. This process can be very time consuming; moreover, certain peripherals can transfer data at a faster rate than can be achieved by the CPU. To overcome this problem, microprocessors have a resource known as **direct memory access** (DMA); this allows data to be transferred directly from a peripheral device to the memory without the intervention of the CPU. The CPU is frozen in a HOLD state while the data transfer takes place.

The control circuitry is fairly involved, but single chip DMA controllers such as the Intel 8257 are available for this type of application.

PROBLEMS

6.1 Discuss the reasons for the use of RAM, ROM, PROM and EPROM in microcomputer systems.

6.2 With the aid of a circuit diagram, show how additional hardware can be used to modify a 1K word RAM having separate input and output lines, to make it appear to the CPU as though it had common I/O lines.

6.3 Draw a schematic diagram showing how a number of 256 x 1-bit RAMs can be interconnected to form a 256 x 8-bit RAM.

6.4 Draw a memory map for a microprocessor using the RAM and ROM arrangement in figure 6.14 with the modification that address lines A_8 and A_{11} are used as inputs to the OR-gate (ROM) and also to the NAND gate (RAM).

7 Input/Output

7.1 Input/Output Ports

Input/output I/O ports are the only method by which the CPU can communicate with devices in the 'real' world. They provide the means not only by which the user can input data to the CPU, but also the means by which the CPU outputs information to peripheral devices. Simplified I/O ports were introduced earlier in the book (see sections 1.5, 3.17 and 4.3); in this chapter we take a more detailed look at their operation together with methods of chip selection.

Early types of I/O port acted simply as an input port or as an output port (the 8212 I/O port is included in this category). Each data line connected to one of these devices is dedicated either as an output line if it is an output port, or as an input line if it is an input port. However, in a range of devices known as **programmable I/O** (PIO) ports, it is possible to 'instruct' each data line associated with the port whether it is to function as an input line or as an output line. For example, a series of instructions can be written into the program which controls the CPU to cause, say, data lines D_0 - D_3 of the PIO to act as input lines and lines D_4 - D_7 to act as output lines.

Additionally, many PIO ports are manufactured with a memory in them. This memory may be either a RAM (for example, the National Semiconductor INS 8154N and the Intel 8155), or a ROM (for example, the Intel 8355) or an EPROM (such as the Intel 8755). The three types of device mentioned above are known as RAM I/O, ROM I/O and EPROM I/O ports, respectively. These devices are extremely versatile and are of great value to the system user.

The data transmission rate which an I/O port may be called on to handle may be very wide, ranging from one pulse every few minutes (or even hours!) in the case of a process control system and up to one-quarter of a million bits per second in the case of a floppy disc memory system. In addition, the I/O port may have to interface to a system requiring an unusual voltage or current level.

Since microcomputers use a common address bus both for memory and I/O transfers, a means must be provided to distinguish between memory transfers and I/O transfers. Three common methods for handling I/O transfers exist, but not all microprocessors utilise all three methods. They are

(1) **Isolated** (or **accumulator**) **I/O**, in which memory locations and I/O addresses are separately decoded (see section 7.2). This method is also known as **standard I/O**.

(2) **Memory-mapped I/O**, in which each I/O port is handled in the same way as a memory location, and each port is allocated an area or location on the memory map (see section 7.3).
(3) **Attached** (or **on-chip**) **I/O**, in which the I/O port is part of the CPU (see section 7.4).

7.2 Isolated I/O or Accumulator I/O Device Selection

One form of isolated I/O arrangement is shown in figure 7.1; this is generally similar to the I/O arrangement in figure 5.1. An input port is selected when the logic signal on the I/O READ control line ($\overline{I/OR}$) is low AND the signal on the

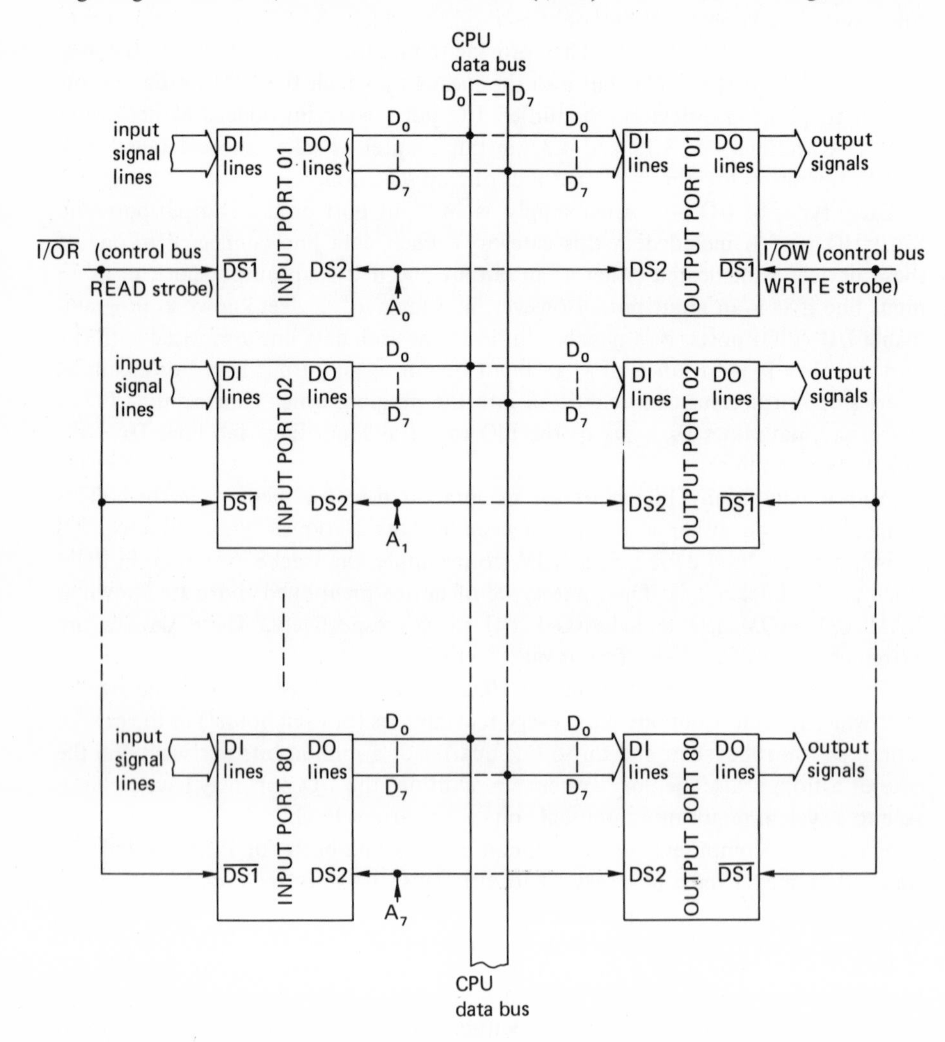

Figure 7.1

address line connected to DS2 is high. Similarly, an output port is selected when the logic signal on the I/O WRITE ($\overline{\mathrm{I/OW}}$) is low AND the signal on the address line connected to the DS2 pin is high. In both the input and the output ports, the port address is established by the signal applied to the DS2 line.

This mode of I/O operation is described as **isolated I/O** addressing because the signal applied to the device selection pin of the I/O device ($\overline{\mathrm{DS1}}$ in figure 7.1) is isolated from or is separate from the signal used for memory selection. That is to say the $\overline{\mathrm{I/OR}}$ and the $\overline{\mathrm{I/OW}}$ signals are used to select input (read) and output (write) devices rather than the $\overline{\mathrm{MR}}$ (memory read signal) and $\overline{\mathrm{MW}}$ (memory write signal). It is also known as **accumulator I/O** since data can only be transferred between an input port and the accumulator, or between the accumulator and an output port. Data transfers between I/O ports and registers other than the accumulator in the CPU are not possible using accumulator I/O. Data transfers using isolated (accumulator) I/O use the INput and OUTput instructions, which were described in section 4.4.

The attention of the reader is now directed to the method of deriving the address of the I/O ports in figure 7.1. As mentioned above, the port address is specified by the address lines used to derive the signal applied to the DS2 control line on the IC (the $\overline{\mathrm{DS1}}$ signal is obtained from the CPU control bus). The reader will note that since only one address line is connected to the DS2 pin on each I/O port, the address decoding is incomplete. As will be seen later, when using only one line to address the I/O chips, the programmer must take great care when writing the program, as the opportunity for creating chaos is enormous!

Referring to figure 7.1, note that for each input port with a given address (say port 01), there is a corresponding output port with the same address. The address of the port is fixed by the address line connected to the DS2 pin of the port, details of this arrangement being described later; what is important is that two ports can apparently have the same address! The reason for this is that input ports are also selected by the $\overline{\mathrm{I/OR}}$ signal, and the output ports by the $\overline{\mathrm{I/OW}}$ signal; consequently, only input port 01 is selected during an INput 01 instruction, and only output port 01 is selected during an OUTput 01 instruction.

From the work in section 4.4, it will be appreciated that the INput and OUTput instructions are two-byte instructions. As a result, only the second byte is available for addressing purposes. That is, an I/O port must have an address in the range 00_{16} to FF_{16}; these addresses are selected using address bus lines A_0 - A_7, giving a maximum number of $2^8 = 256$ I/O addresses.

Using the simple addressing arrangement in figure 7.1, where only one address line is connected to the DS2 pin of the IC, then that line must be driven high in order to select the port. In the case of figure 7.1, the user can address a maximum of eight input ports and eight output ports (this is because each port is connected to one of the address lines A_0 - A_7). This arrangement is adequate to meet the needs of many simple systems. Thus, when the signal on address line A_0 is high, then either input port 01 or output port 01 is selected (this depends on whether an IN or an OUT instruction is currently being executed). When the signal on the address bus line A_1 is high, then either input port 02 or output port 02 is selected. Similarly, when address line A_7 is high, then either input port 80 or output port 80 is selected. The signals applied to the address bus lines to select the I/O ports in figure 7.1 are summarised in table 7.1.

Table 7.1

Signal on address lines A_7 - A_0								Hex signal on address bus	Address of I/O port
A_7	A_6	A_5	A_4	A_3	A_2	A_1	A_0		
0	0	0	0	0	0	0	1	01	01
0	0	0	0	0	0	1	0	02	02
0	0	0	0	0	1	0	0	04	04
0	0	0	0	1	0	0	0	08	08
0	0	0	1	0	0	0	0	10	10
0	0	1	0	0	0	0	0	20	20
0	1	0	0	0	0	0	0	40	40
1	0	0	0	0	0	0	0	80	80

If the address lines A_0 - A_7, inclusive, are fully decoded, then 256 input ports and 256 output ports can be selected. Examples of generating active-high chip-select signals are illustrated in figure 7.2. Consider for the moment the case where an active-high signal must be generated in order to address I/O port 13_{16}. Using

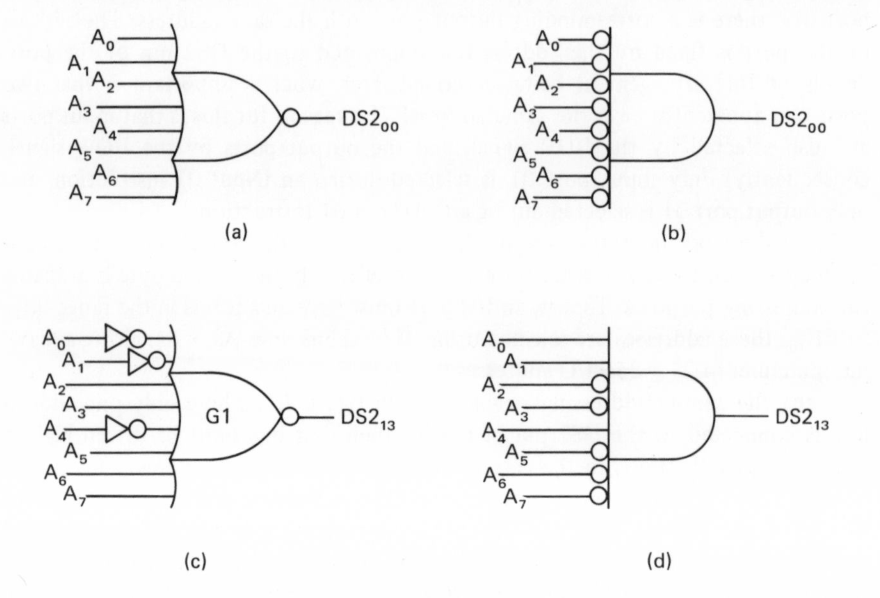

Figure 7.2

NOR gate G1 (figure 7.2c), all inputs to the gate must be low to force the output signal $DS2_{13}$ high; since the signals on the address bus lines A_0, A_1 and A_4 are inverted, then the signals on these address bus lines must be high to satisfy the above condition. The binary signals which must be output on the lower byte of the address bus in order to select I/O port 13_{16} are as follows (A_7 first and A_0 last): 00010011_2. The arrangement in figure 7.2d can also be used to select I/O port 13_{16}. The logic circuit in figure 7.2a and b can be used to uniquely provide an active high signal for I/O port 00_{16}.

A limitation of the method of address decoding in figure 7.2 is that a multi-input gate must be used for each I/O address. One method of overcoming this limitation is by the use of IC decoders, as shown in figure 7.3. The decoders used in the figure produce one-of-eight active low output signals (a typical IC is the 8205 - see section 3.16). If an active high signal is required (such as for the DS2 signal in figure 7.1), an invertor is needed between the appropriate output from the decoder and the device select pin on the I/O port. Address lines A_0 through A_2 are used to address one of the eight output lines of each decoder, whilst address bus lines A_3 and A_4 together with an appropriate logic signal level are used to enable the decoders; each active low output signal is connected to a $\overline{\text{device select}}$ pin (or via an invertor to an active high DS pin) of an I/O port. With the connections shown in figure 7.3, twenty-four I/O ports (ports 00_{16} - 17_{16}) can be selected.

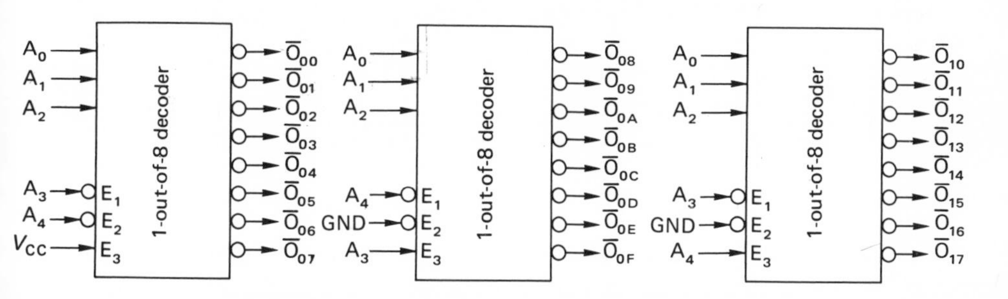

Figure 7.3

Features of isolated I/O

Many processors provide control signals which allow the user to distinguish between I/O operations and memory read/write operations. Included in this group are the Intel 8080 and 8085, the Zilog Z80 and the Signetics 2650.

Advantages of isolated I/O

(1) I/O and memory systems can be separated from one another.
(2) I/O port addresses can be simple, leading to simple decoding.
(3) Programs are easy to interpret, since I/O and memory transfers are easily distinguished from one another.

Disadvantage of isolated I/O

Data transfers must either be to or from the accumulator, that is, data cannot be transferred into or from an auxiliary register.

7.3 Memory-mapped I/O

Where memory-mapped I/O is used, either the memory read strobe ($\overline{MR}$) or the memory write strobe ($\overline{MW}$) is used in I/O chip selection. That is to say, each output port is regarded as a memory location and is allocated an area on the memory map of the system. Certain microprocessors can only address I/O ports by this means, including the Motorola 6800, and the National 8060 and PACE. CPUs which can use isolated I/O can also employ memory-mapped I/O.

A block diagram illustrating a pair of memory-mapped I/O ports is illustrated in figure 7.4. The reader will note that the $\overline{DS1}$ pin of the input port is activated by the memory read strobe ($\overline{MR}$), and $\overline{DS1}$ on the output port is energised by the memory write strobe ($\overline{MW}$). The address select signal (DS2) is derived from address lines A_9 and A_{10} via gates G1 and G2, to form the $(A_9.\overline{A}_{10})$ signal. When the latter signal is high AND the $\overline{MR}$ strobe is low, the input port is enabled. When the $(A_9.\overline{A}_{10})$ signal is high and the $\overline{MW}$ strobe is low, the output port is enabled.

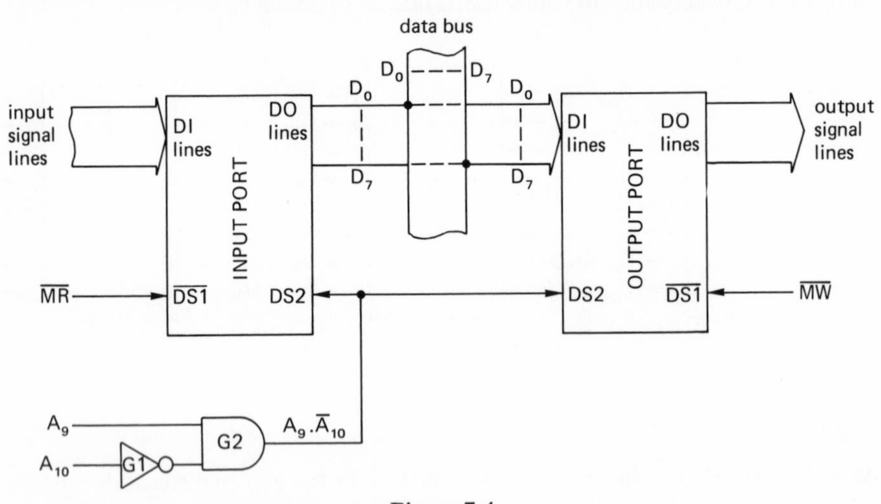

Figure 7.4

Since only two address lines are used in the address decoding circuit, both I/O chips have many addresses. These addresses are determined from the address combinations below

A_{15} A_{14} A_{13} A_{12}	A_{11} A_{10} A_9 A_8	A_7 A_6 A_5 A_4	A_3 A_2 A_1 A_0
X X X X	X 0 1 X	X X X X	X X X X
$0\text{–}F_{16}$	$2, 3, A, B_{16}$	$0\text{–}F_{16}$	$0\text{–}F_{16}$

where X = '0' or '1'. The above addresses are determined by the fact that address lines A_0 through A_8 and A_{11} through A_{15} are not used, and each can have a logic '0' or a logic '1' on them. The logic levels on the DI lines of the input port are read by the CPU when the program calls for a **memory read** from one of the addresses including 0200, 0201, 0202 . . . 02FF, or 1200, 1201 . . . 12FF, or F200 . . . F2FF, or 0300 . . . 03FF, etc.

The above decoding was chosen to fit into the memory map in figure 6.15 (which corresponds to the circuit decoding in figure 6.14). The memory map for the first 4K words of memory is shown in figure 7.5; this map is repeated in each 4K word group up to the maximum addressing capability of 64K words. That is

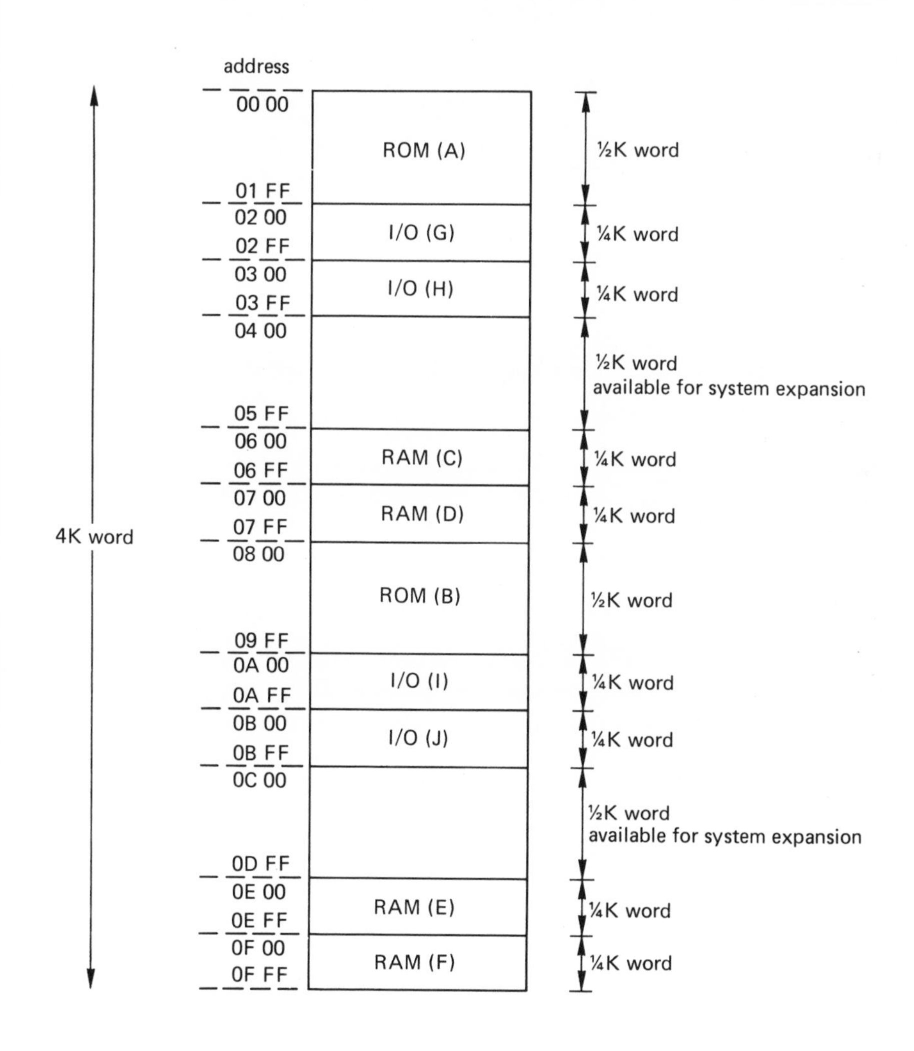

Figure 7.5

to say, each input port or each output port can be addressed in the first 4K locations by any address in the range 0200 - 03FF or 0A00 - OBFF; these addresses are repeated sixteen times in the 64K locations.

The reader will appreciate that the above method of addressing is expensive in terms of usage of area of the memory map, and is the cost which must be paid when using simple decoding methods. More complex decoding methods lead to more efficient utilisation of the area available on the memory map.

Advantages of memory-mapped I/O

(1) Any instruction which operates on data in a memory location can be used in conjunction with data from an I/O port.
(2) Separate I/O instructions, that is, IN and OUT, are unnecessary.
(3) A wider range of addresses is available when compared with isolated I/O.
(4) I/O port addresses can be decoded by the same equipment used for memory address decoding.

Disadvantages of memory-mapped I/O

(1) I/O ports occupy areas on the memory map.
(2) It is less suitable for small and/or simple systems.
(3) It is difficult to distinguish in the program between memory and I/O data transfers.

General note: System designers fall into two groups, namely either 'isolated I/O people' or 'memory-mapped people'. Both will argue the superior merits of their approach!

7.4 Attached I/O or on-chip I/O

A number of CPUs have I/O facilities on the CPU chip itself; included in these are the Fairchild F-8, the Intel 8048 and the National 8060. The cost paid for this feature is that a number of the CPU connecting pins must be dedicated to I/O operations. CPUs incorporating this facility are generally aimed at systems needing only a small number of support chips, that is, at systems which have a low 'chip count'. I/O facilities are also available on a number of memory chips (which may either be RAM, or ROM or EPROM).

7.5 Handshaking

Handshaking is the name given to the technique in which the CPU and the peripheral 'tell' each other (via I/O ports) that data is ready to be transferred. There are two types of handshake, namely an input handshake and an output handshake.

A handshake system involves the use of three ports (see, for example, figure

7.6), two of them (one an input port and one an output port) each dealing with a single handshake line; consequently, these ports could each be a single flip-flop or one line of an 8-line I/O port. The remaining port is required for the transmission of information between the peripheral and the CPU. If a programmable I/O port is used, the requirement may be reduced to one port simply by programming two of the lines to act as handshake lines, the remaining six lines being available for data transmission purposes.

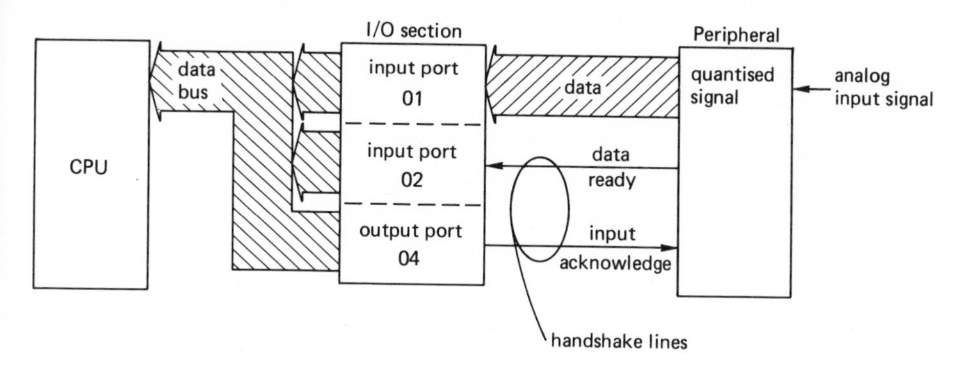

Figure 7.6

The basis of an **input handshake** is shown in figure 7.6. Suppose that the peripheral in the figure is an analog-to-digital convertor (ADC), in which successive analog signals are to be quantised by the ADC (an example of this type is dealt with in chapter 11), the quantised signals being stored in successive locations in RAM. In the following it is assumed that isolated I/O port selection has been adopted, and that the addresses of the ports in figure 7.6 are 01, 02 and 04.

The ADC is a device which converts an analog voltage into a digital value (in this case we assume it is converted into an 8-bit word), but it takes a finite time to complete the process. At the commencement of the quantisation process, the QUANTISED SIGNAL at the output terminals of the ADC is reset to zero; as the quantization process continues, so the binary value at these terminals increases until, finally, the analog input voltage is converted into its equivalent binary value. During the above period of time it is essential to prevent data being transmitted to the CPU. Once quantisation is complete, the CPU is informed that data is ready; the CPU then initiates the data reading process. The general procedure is as follows

(1) The peripheral sends a DATA READY signal to input port 02, which is read by the CPU. Until this signal is received, the CPU either remains in a software 'waiting loop' or continues with some other part of its program.

(2) The CPU reads the QUANTISED SIGNAL from input port 01.

(3) The CPU sends an INPUT ACKNOWLEDGE signal via port 04 to acknowledge receipt of data. In the case of an ADC, the INPUT ACKNOWLEDGE signal

is used to initiate the next analog-to-digital conversion process; this completes the input handshake between the peripheral and the CPU.

An **output handshake** system is shown in figure 7.7; the peripheral could be, for example, a teletype or a printer. The general procedure adopted is as follows

(1) The CPU sends an OUTPUT READY signal to the peripheral via output port 40.

(2) When the peripheral is ready to receive data, it sends a PERIPHERAL READY signal to input port 20, which is read by the CPU.

(3) The CPU sends data to the peripheral via output port 10. Step 1 may be repeated when the peripheral completes its operation; this completes the output handshake between the peripheral and the CPU.

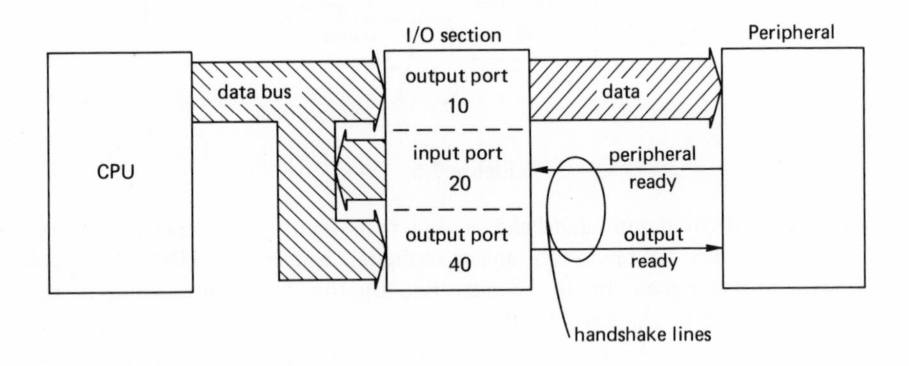

Figure 7.7

7.6 A Programmable I/O Port containing a Memory

An LSI chip containing I/O ports together with a small amount of memory provides the user with a very useful range of facilities. The user can define, by means of software, each line (or group of data lines) associated with the port either as an input line or as an output line. Thus, if the system requires the use of four input lines and four output lines, the user can specify in his program which four lines he wishes to use as input lines and which four as output lines. Moreover, many of these devices are manufactured in a 40-pin DIL package, so that two (sometimes three) I/O ports are contained in the same package. Consequently, a single 40-pin DIL provides the user with a small amount of memory with sixteen programmable I/O lines (assuming, that is, that it contains two I/O ports).

The memory contained on the chip may be RAM (for example, the Intel 8155/56, or the National INS 8154) or ROM (for example, the Intel 8355) or EPROM (for example, the Intel 8755).

7.7 Organisation of a Typical Programmable RAM I/O Port

A block diagram of one form of programmable I/O chip containing 128 bytes of RAM is shown in figure 7.8. The I/O section contains two ports, port A and port B, each capable of handling eight I/O lines. Since each line may be programmed either as an input line or as an output line, data transfer with handshake can be arranged. The RAM section contains 1K bits of RAM, organised in the form of 128 x 8 bits ($\frac{1}{8}$K words).

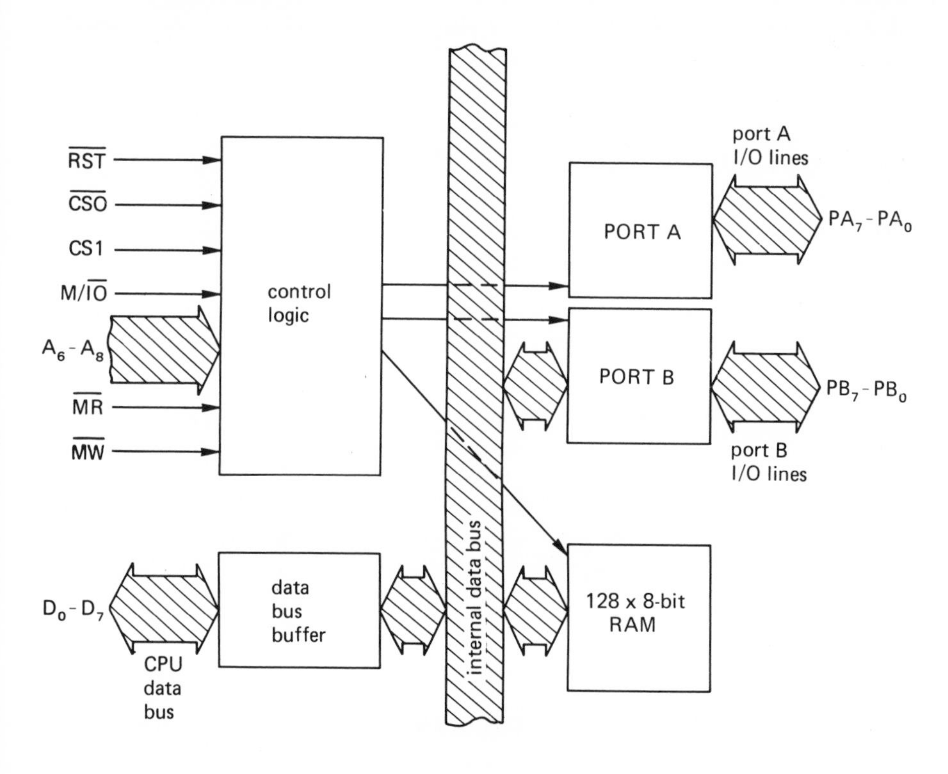

Figure 7.8

Each I/O port has its own I/O connecting lines (port A I/O lines are PA_7 - PA_0, and port B lines are PB_7 - PB_0), and the bidirectional data bus from the CPU is connected via the data bus buffer to the internal bus of the chip. The operations within the chip are governed by the control logic section, whose input lines are described below.

$\overline{RST}$ (reset) When the reset signal is low, lines PA_7 - PA_0 and PB_7 - PB_0 are forced to their high impedance state, and all operations within the RAM I/O chip are inhibited. Immediately following the reset pulse, all the lines of both I/O ports are set to the 'input' mode and the handshake facilities are deselected. The contents of the RAM are unaffected.

$\overline{CS0}$ and CS1 (chip select) The address lines connected to these pins fix the address of the device on the memory map.
M/$\overline{IO}$ (memory, I/O select) When this pin is at logic '1' the RAM section of the chip is selected; when at logic '0', the I/O section is selected.
A_6 - A_0 (address inputs) these address lines are used in association with the 128 byte RAM (remember, $2^7 = 128$), and define an address within the RAM.
$\overline{MR}$, $\overline{MW}$ (read and write strobes - active low) Since the chip contains memory, it must be allocated an area on the memory map. Consequently, the strobe signals used in connection with the chip are the memory read strobe ($\overline{MR}$) and the memory write strobe ($\overline{MW}$). A logic '0' on the $\overline{MR}$ line instructs the chip that data is being read either from the RAM or from a port; data are read from the RAM if, at this time, the signal on the M/$\overline{IO}$ pin is high, and from a port if it is low. A logic '0' on the $\overline{MW}$ line instructs the chip that data is being written either to RAM or to a port.

7.8 An Application of a Programmable I/O Port

Figure 7.9 shows a simplified block diagram of a programmable I/O (PIO) chip of the type in figure 7.8. For simplicity, the internal structure has been omitted.

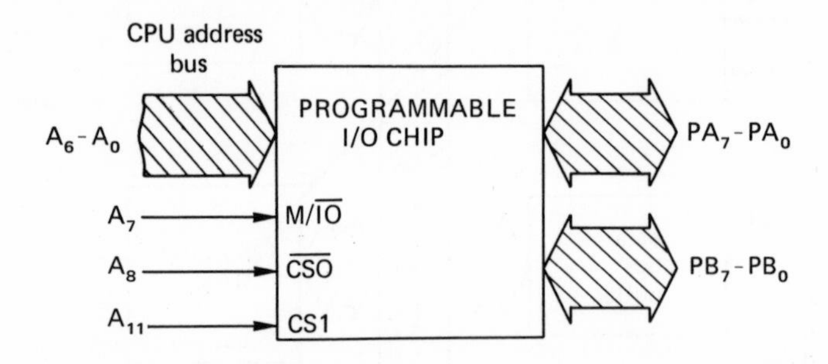

Figure 7.9

Address lines A_6 - A_0 of the CPU address bus are needed for the purpose of addressing the 128 bytes of data in the RAM (see section 7.7). The remaining address bus lines (A_{15} - A_8) are therefore available for chip selection. In figure 7.9, the chip is selected when A_8 is low AND A_{11} is simultaneously high. Since address lines A_9, A_{10} and A_{12} through A_{15} are not used, the address of the PIO in figure 7.9 is given by

A_{15}	A_{14}	A_{13}	A_{12}	A_{11}	A_{10}	A_9	A_8	A_7	A_6	A_5	A_4	A_3	A_2	A_1	A_0
X	X	X	X	1	X	X	0	X	X	X	X	X	X	X	X
0–F_{16}				8, A, C, E_{16}				used in memory/IO addressing							

That is, the PIO chip is chosen by any one of the addresses X8XX, XAXX, XCXX, XEXX, where X is a hexadecimal value in the range 0 to F. In the following we use the range of addresses 0EXX when dealing with the PIO; this is an arbitrary choice, and any other range such as 88XX could be used.

Referring to figure 7.9, the reader will observe that the RAM section of the chip is selected when address line A_7 is high; the I/O section is selected when A_7 is low. Thus, when the program addresses location $0E80_{16}$, then it addresses location 'zero' in the RAM section of the RAM I/O chip; address $0EFF_{16}$ selects the final (the 128th) location in RAM. Any address in the range 0E00 to 0E7F results in the address bus line A_7 having a logic '0' on it, so that this range of addresses refer to the I/O section of the chip.

The use of a PIO is dealt with in the following application. Suppose that it is necessary to devise a system which monitors the state of the signals from four switches, and outputs the state of these switches to four lamps. That is to say, if the signal from a certain switch is logic '1', then the appropriate lamp must be illuminated; if the signal from the switch is logic '0', then the lamp must be extinguished. Moreover, after a short time delay, the CPU must output the state of the switches to another set of four lamps.

The general arrangement of the PIO port together with the switches and lamps are shown in figure 7.10; the address and chip select lines are identical to those

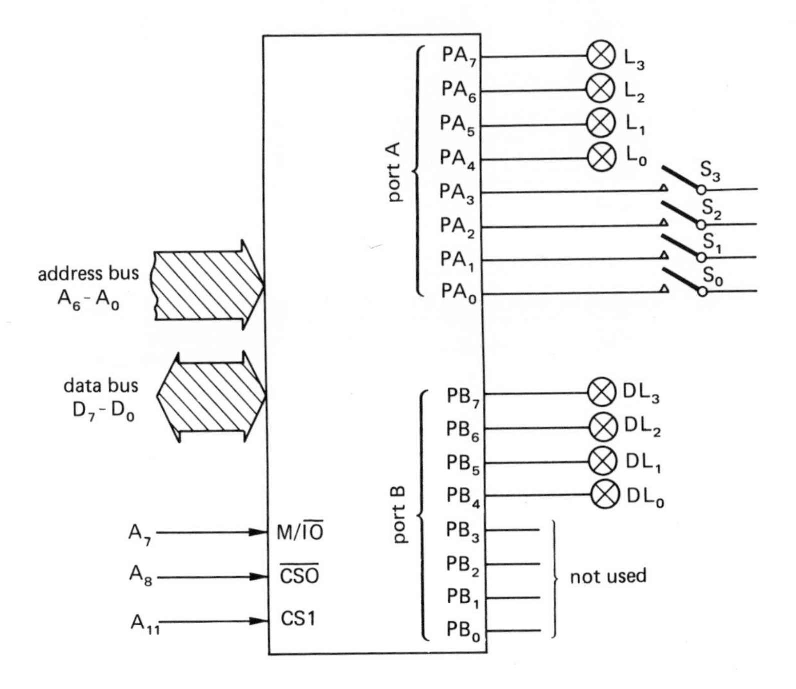

Figure 7.10

used in figure 7.9. Switches S_3 - S_0 are connected to lines PA_3 - PA_0, respectively, while the lamps L_3 - L_0 (which are to be illuminated instantly (or nearly so) by the operation of the switches) are connected to port A lines PA_7 - PA_4, respectively. The program (see table 7.3) is written so that switch S_0 controls lamp L_0, switch S_1 controls lamp L_1, etc. Thus lines PA_3 - PA_0 are defined as input lines, and lines PA_7 - PA_4 as output lines.

After a suitable time delay, the state of each switch is to be indicated on lamps DL_3 - DL_0; lamp DL_3 gives a delayed indication of the state of switch S_3, DL_2 giving a delayed indication of the state of switch S_2, etc. As a result, lines PB_7 - PB_4 must also be defined as output lines. Since lines PB_3 - PB_0 are not used, they may either be defined as input lines or as output lines; in this case (see table 7.3) they are defined as input lines.

Figure 7.11 shows the functions allocated to certain locations within the PIO chip. The lowest available address is 0E00 (note: with the chip selection signals used, this could equally well be 0800, 0A00 or 0C00). The first thirty-two addresses

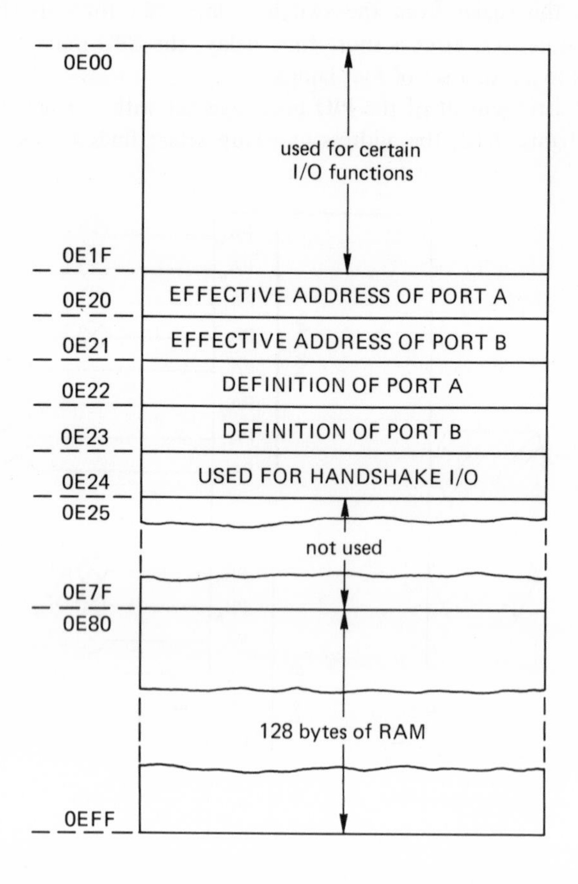

Figure 7.11

(0E00 to 0E1F) are used for I/O functions which are not of interest to us in this application. Address 0E20 is the effective address of port A, and 0E21 is the effective address of port B. Thus if we wish to output data to port B, the program must call for data to be transferred to memory location 0E21 (remember, memory-mapped I/O is used in this case). Location 0E22 is a register which, under program control contains data defining the operation of port A (that is, the information in this register must define PA_7 - PA_0 as output lines and PA_3 - PA_0 as input lines). Location 0E23 is another register containing data defining the operation of port B. Location 0E24 is used in association with handshaking I/O operation and is not described here.

To define the operation of port A, a binary word is written under the control of the program into location 0E22 (see figure 7.11). To define the operation of port B, another word is written into location 0E23. In the PIO port described here, when a logic '1' is written into a location in the port definition word, it causes the appropriate output line (see below) to function as an output line. When a logic '0' is written into a location in the port definition word, it causes the appropriate output line to function as an input line. For example, if the binary word 11000011_2 ($C3_{16}$) is written into location 0E22 in the PIO then, since bit 7 = '1', line PA_7 of port A is an output line; applying this argument to each bit in the word, lines PA_7, PA_6, PA_1 and PA_0 are output lines. Since bits 5, 4, 3 and 2 of the port defining word have logic 0's in them, lines PA_5, PA_4, PA_3 and PA_2 are defined as input lines.

If the binary word 00111100_2 ($3C_{16}$) is written into location 0E23, then lines PB_7, PB_6, PB_1 and PB_0 of port B become input lines, and lines PB_5, PB_4, PB_3 and PB_2 are output lines.

Before commencing with a detailed study of the program, refer to table 7.2 which outlines the operations carried out in association with each instruction in the program.

Table 7.2

Instruction	Op-code	Operation performed
MVI A,data	3E	(A) ← DATA
STA address	32	(ADDR) ← (A)
LDA address	3A	(A) ← (ADDR)
RLC	07	rotate accumulator contents left one bit
MOV B,A	47	(B) ← (A)
MOV A,B	78	(A) ← (B)
JMP	C3	(PC) ← (ADDR)

The two-byte MVI A,$F0_{16}$ instruction in addresses 0F20 and 0F21 in table 7.3 causes the data ($F0_{16}$ or 11110000_2) in the second byte to be MoVed Immediately

Table 7.3

Memory address (hex)	Memory contents (hex)	Instruction (mnemonic)	Comment
0F20	3E	MVI A,$F0_{16}$	
21	F0		
22	32	STA 0E22	Define port A
23	22		
24	0E		
25	32	STA 0E23	Define port B
26	23		
27	0E		
28	3A	LDA 0E20	'Read' switches
29	20		
2A	0E		
2B	07	RLC	Shift (A) four places left
2C	07	RLC	
2D	07	RLC	
2E	07	RLC	
2F	32	STA 0E20	Illuminate L_3 - L_0
30	20		
31	0E		
32	47	MOV B,A	Save (A)
33			Delay program
.			
.			
.			
3E			
3F	78	MOV A,B	Restore (A)
40	32	STA 0E21	Illuminate DL_3 - DL_0
41	21		
42	0E		
43	C3	JMP 0F28	Return to effective start of program
44	28		
0F45	0F		

into the Accumulator. Referring to table 7.2, this operation is described as

$$(A) \leftarrow \text{DATA}$$

That is, DATA (a numerical value) is transferred into the accumulator. The parentheses around letter A are a shorthand way of saying "the contents of" the accumulator.

The three-byte instruction in locations 0F22 - 0F24 causes the CPU to STore the Accumulator (STA) contents in the memory address given in the next two bytes of the instruction. The location specified is 0E22, which is the address associated with the port A definition word. This causes the word $F0_{16}$ to be moved into 0E22, which results in port A lines PA_7 - PA_4 being defined as output lines, and PA_3 - PA_0 as input lines (see also the external connections to the PIO chip in figure 7.10). The instructions in locations 0F25 - 0F27 also cause the same definition word to be moved into address 0E23; that is, lines PB_7 - PB_4 also become output lines, and PB_3 - PB_0 become input lines (although the latter are not used in this case). The three-byte instruction in locations 0F28 to 0F2A causes the CPU to LoaD the Accumulator (LDA) from the memory address specified in the second and third bytes of the instruction; that is, the accumulator is loaded with the contents of address 0E20. Referring to figure 7.11, the reader will see that the effect of this instruction is to load the accumulator with data from port A. Thus the state of switch S_3 (see figure 7.10) is loaded into bit 3 of the accumulator, the state of switch S_2 is loaded into bit 2 and so on. On the completion of this instruction, the least significant four bits of the accumulator store the states of the four switches connected to the 'input' lines of port A.

The instruction at program address 0F2B is a one-byte instruction which causes the CPU to Rotate the contents of the accumulator Left Circular (RLC) one bit; at the same time the condition of the 'carry' status flag may alter, but this does not affect the operation of the program and this effect is ignored. The effect of one RLC instruction is illustrated in figure 7.12. The reader will note that each bit in the accumulator is shifted left; that is the content of b_0 is shifted to b_1, the content of b_1 is shifted to b_2, etc., and the content of b_7 is rotated into b_0. At the same time, the bit in b_7 is copied into the carry status flag (C); the carry flag may store either a '1' or a '0' before the RLC instruction, but after the instruction has been executed it stores a copy of the data which was in b_7, that is, a logic '0'.

The program calls for this instruction to be repeated three times more, so that ultimately the state of b_3 is transferred to b_7, b_2 to b_6, b_1 to b_5, and b_0 to b_4. That is, the signal from switch S_3 is now stored in b_7 of the accumulator, the condition of S_2 is stored in b_6, the state of S_1 is stored in b_5, and S_0 is stored in b_4.

The 3-byte instruction at addresses 0F2F - 0F31 (STA 0E20) cause the contents of the accumulator to be transferred (stored) in address 0E20. Reference to figure 7.11 shows that this causes the contents of the accumulator to be transferred to port A. Since lines PA_7 - PA_4 are defined as output lines, the data in b_7 of the accumulator is transferred via line PA_7 to lamp L_3; similarly, the logic signal applied to lamp L_2 corresponds to the binary value in b_6 of the accumulator, etc. Since the data transfers between the microprocessor and the port occurs in a very short time, the output port must latch the data if a continuous display is required.

Having transmitted the condition of switches S_3 - S_0 to lamps L_3 - L_0, respectively, it is necessary to 'save' the contents of the accumulator. The reason for doing this is that the program specification calls for a time delay between illumi-

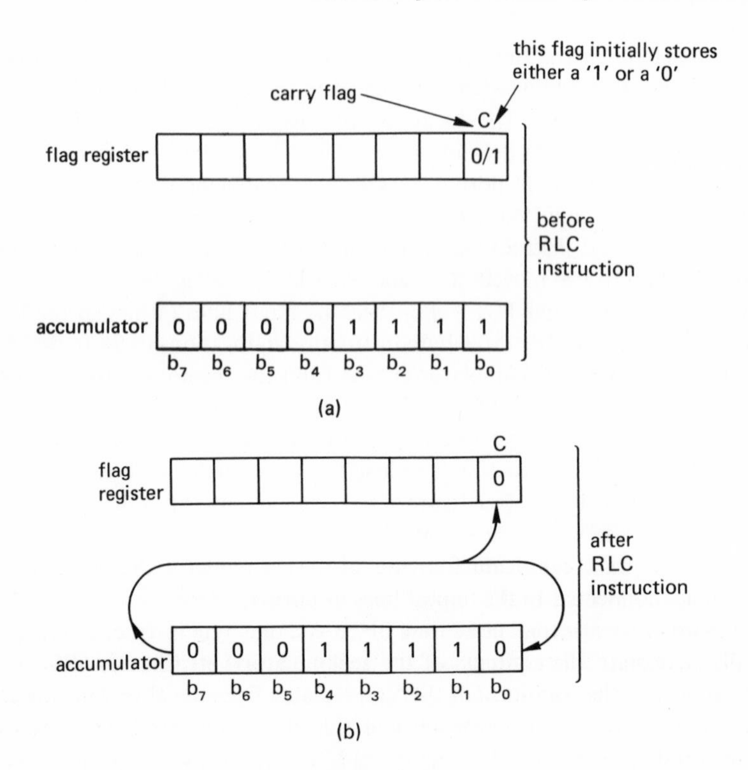

Figure 7.12

nating lamps L_3 - L_0 and illuminating lamps DL_3 - DL_0 (see figure 7.10). During the 'time delay' program, the accumulator is used, and its contents will almost certainly change in value. The accumulator contents can be saved or stored in any one of a number of ways, including the use of a 3-byte STA instruction. In this case, a 1-byte MOVe instruction is used; this allows the CPU to move data between registers or between a register and a memory location (or vice versa). The reader will recall that the CPU contains a number of auxiliary registers, and here we use an instruction in address 0F32 of the program which MOVes the contents of the accumulator into register B.

The instructions in locations 0F33 to 0F3E, inclusive, implement a time delay program or 'do nothing' program; the function of this section of the program is to cause the microcomputer to 'mark time' for a fixed period of time before executing the next section of the program. The details of the time delay program are omitted at this time for simplicity, but the reader who wishes to know more about this section of the program is referred to chapter 8 (section 8.5), where a time delay program is described. Some microcomputers have a time delay instruction in their instruction set and others do not. In the latter case, the programmer must write a program which causes the CPU to execute a number of time wasting instructions.

After an appropriate time delay, the CPU executes the MOVe A,B instruction at address 0F3F, which moves the contents of register B into the accumulator, thus restoring the contents of the accumulator to its condition before instruction 0F32. The 3-byte STA 0E21 instruction results in the delayed data being transmitted to lamps DL_3 - DL_0 via port B. Finally, an unconditional jump is made to instruction 0F28, at which point the state of the switches are 'read' once more and the loop recommenced.

In the above, the program was stored in a RAM (or in a ROM) whose memory map addressing included the page 0F00 - 0FFF. The program can also be written into the RAM section of the RAM/IO chip as follows. Since the first RAM location of the RAM/IO device is at 0E80 (see figure 7.10 and associated text), the corresponding commencement address of the program in table 7.2 is 0EA0 (that is, at address $0E80_{16} + 20_{16}$) and the final address is at 0EC5 (that is, $0E80_{16} + 45_{16}$).

7.9 Serial Data Transmission

So far it has been the case that eight bits of data are transmitted simultaneously along the eight data bus lines. This is known as **parallel data transmission.** However, many computer peripherals transmit data as a series of on-off pulses along a single wire. An example of this is digital data transmission over a telephone wire. This is known as **serial data transmission**; peripherals using this mode of data transmission include teletypes, CRT computer terminals, and modems (MODulator/DEModulatorS).

One method of interfacing between a microprocessor and a serial data transmission system is by the use of dedicated input and output lines on the CPU chip. On the Intel 8085 chip, these are the SID line (Serial Input Data) and the SOD line (Serial Output Data). Special instructions are included in the instruction set to cause a bit to be input from the SID line to a register (which may either be the accumulator or an auxiliary register), or to be output to the SOD line. This technique is sometimes colloquially known as 'bit banging'. A series of instructions causes data either to be serially input from the incoming line, or to be serially output to the line.

Another, and more popular method, is to use dedicated hardware devices in association with the CPU; these devices convert between the serial incoming or outgoing data and the parallel data mode used by the CPU bus system. A popular device for this application is the UART (Universal Asynchronous Receiver/Transmitter).

Before describing a UART, the general format of a serial data signal is outlined. A serial 'character' (which may represent a number or a letter of the alphabet, or some other character) consists of a consecutive group of bits comprising the following:

(1) A start bit (logic '0')
(2) Five to eight data bits (either logic '0' or '1')

(3) A parity check bit ('0' or '1') which is determined from the data bits (either odd or even parity may be used).
(4) One, one and a half, or two stop bits (logic 1's)

In addition, there may be a number of 'idling bits' (logic 1's) between characters.

An example of a serial character is shown in figure 7.13a, which contains a start bit, eight data bits, a parity bit and a stop bit. A fundamental feature of serial data transmission is that the data can be transmitted asynchronously; that is the data can be transmitted in bursts ranging from a maximum rate when one character after another is transmitted, to a much slower rate with 'idling' gaps between characters. An example of the latter occurs where a video display terminal is used; during a period of time when the computer transmits a message to the CRT for display purposes, it does so at the maximum rate, but there are other periods when the computer stops transmitting data and generates only idling bits. One function of the UART is to detect the completion of one character and the beginning of the next, which it does at the logic '1' to '0' change between the stop bit (or the final idling bit) and the start bit. An example of data being transmitted at less than the maximum rate is illustrated in figure 7.13b.

The serial data is said to be 'framed' by the start and stop bits, and another function of the UART is to look for 'framing' errors – that is, it checks if it has received a character without the correct number of stop bits. Likewise, the UART checks the received word for correct parity.

Figure 7.13(b) illustrates different rates of data transmission. The nth and $(n + 1)$th characters are transmitted at the maximum rate, with the stop bit of the nth character being followed by the start bit of the $(n + 1)$th character. The $(n + 1)$th and $(n + 2)$th characters are transmitted at less than the maximum rate, with idling bits being transmitted between the stop and start bits of the respective characters.

Each character in figure 7.13 has eleven bit 'slots' or 'bit-times' (comprising a start bit, eight data bits, a parity bit and a stop bit). The maximum transmission rate is known as the **baud rate**, where

$$\text{baud rate} = (\text{No. bit-times per character}) \times (\text{maximum character transmission rate})$$

If the maximum data transmission rate is 10 characters per second, then

$$\text{baud rate} = 11 \times 10 = 110 \text{ baud}$$

the **bit time** is given by

$$\text{bit time} = \frac{1}{\text{baud rate}} = \frac{1}{110} \text{ s} = 9.09 \text{ ms}$$

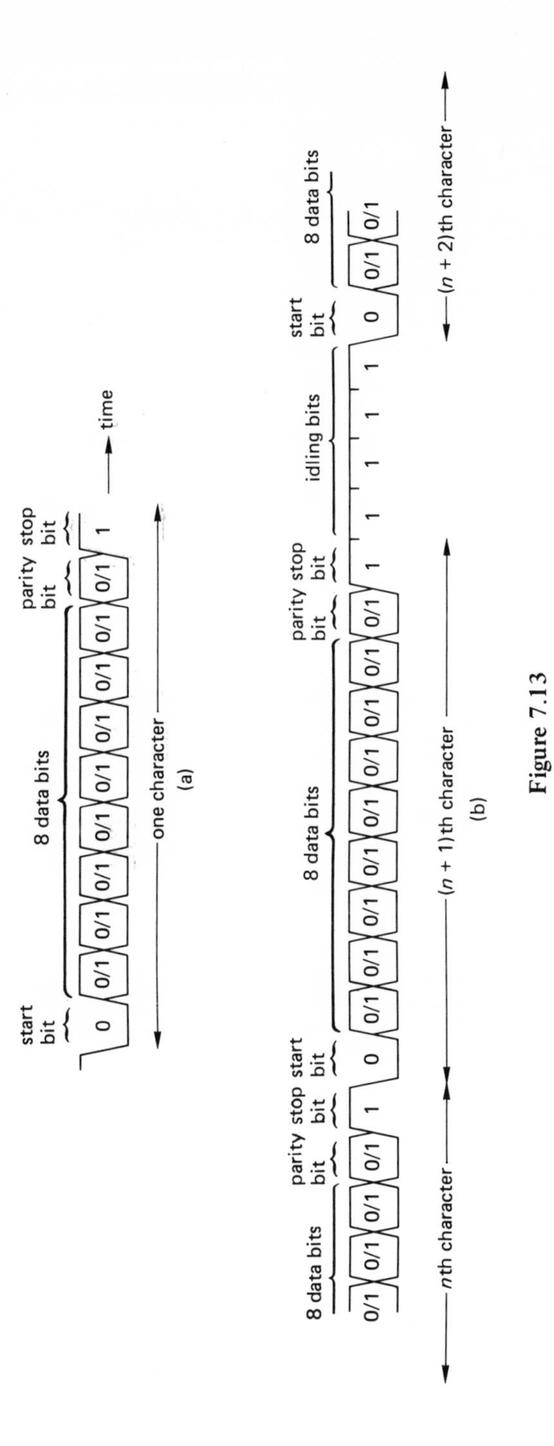
start bit
8 data bits
parity bit
stop bit
0
0/1
1
time
one character
(a)
8 data bits
parity bit
stop bit
start bit
idling bits
nth character
(n + 1)th character
(n + 2)th character
(b)

Figure 7.13

Sometimes the **data rate** is quoted, where

data rate = (No. data bits per character) x (maximum character transmission rate)

In the above case

$$\text{data rate} = 8 \times 10 = 80 \text{ bits/s}$$

A simplified block diagram of a UART is shown in figure 7.14. Although a UART is a single IC, it can be regarded as though it contains separate receiver and transmitter sections. The receiver section contains a shift register capable of storing an incoming character, the data being shifted through the register under the control of a series of pulses applied to the receiver clock line. When a complete character has been received, it is checked for a number of features including parity, framing error and overrun (which occurs when the UART receives a new character before the previous one was 'read'). After receiving a correct character, a signal is transmitted on the RXA line (received data available) to the CPU; when the CPU receives the RXA signal, it causes the data in the receiver register to be transferred to the CPU data bus.

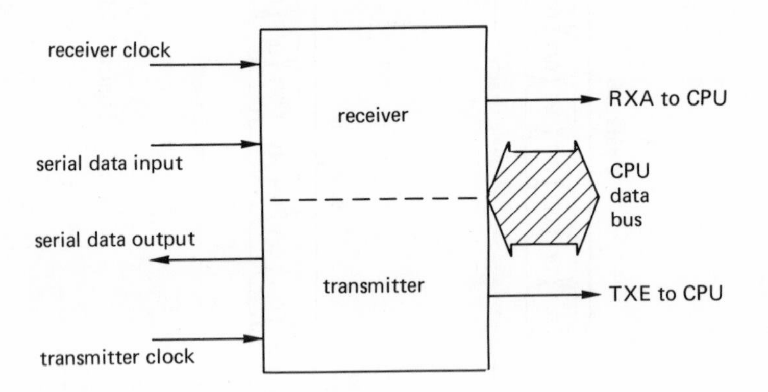

Figure 7.14

When transmitting data, the transmitter section of the UART sends a signal to the CPU via the TXE line (transmitter buffer empty) that the previous character has been transmitted, and that it is ready to receive new data. The CPU then transfers eight bits into the transmitter register. When the start, parity and stop bits have been added, the data is shifted serially out of the UART. If no further data is forthcoming from the CPU, the transmitting line is driven high (corresponding to the idling condition) until the beginning of the next character (which is marked by a start bit - logic '0').

Since a UART contains both a receiver and a transmitter, it can receive data from one source and transmit data to another source. Moreover, since the receiver and transmitter have their own clock signals, the two sections of the UART can operate at differing baud rates. However, the data format used by both halves of the UART must be the same. The baud rate of the UART clock signals can either be generated by dividing down the microprocessor clock frequency to give an appropriate frequency or, alternatively, a programmable bit-rate generator can be used whose frequency is controlled by the CPU via an output port.

7.10 Using Shift Registers as Serial I/O Devices

A shift register is simply a register in which the stored data can be shifted either to the right or to the left under the control of a clock pulse. One clock pulse signal causes the data in the register to be shifted by one stage (or step) through the register.

The block diagram in figure 7.15 shows one method of using a number of shift registers in conjunction with an I/O port to expand the I/O capability of a microcomputer system. Each line of output port A applies data to the INPUT line of a shift register. The signal applied to the CLK (clock) line causes the data in the register to be moved one step right along the register. Hence four CLK pulses cause the data which was initially at DO_0 to be moved along the register to output Z_0.

The network in figure 7.15 can supply data to up to eight shift registers. In the case considered these are 4-bit registers, so that up to 32 output signals can be controlled.

At the instant of switch-on, the user has no guarantee that the outputs of the registers will be zero. It may therefore be necessary to apply a CLR (clear) signal via output port B to the shift registers at the outset of the program. Alternatively, one of the lines from port A (say DO_7) could be used to activate the CLR line.

7.11 Interface Standards

The interconnections (the **interface**) between the microcomputer and peripherals presents many problems. To minimuse these problems a number of interface standards have been developed. The three primary standards are

(1) The IEEE-488 interface
(2) The RS-232 interface
(3) The S-100 interface

The IEEE-488 interface

This was first developed by the Hewlett-Packard Company for the interconnection of programmable instruments, and was adopted as a standard by the American

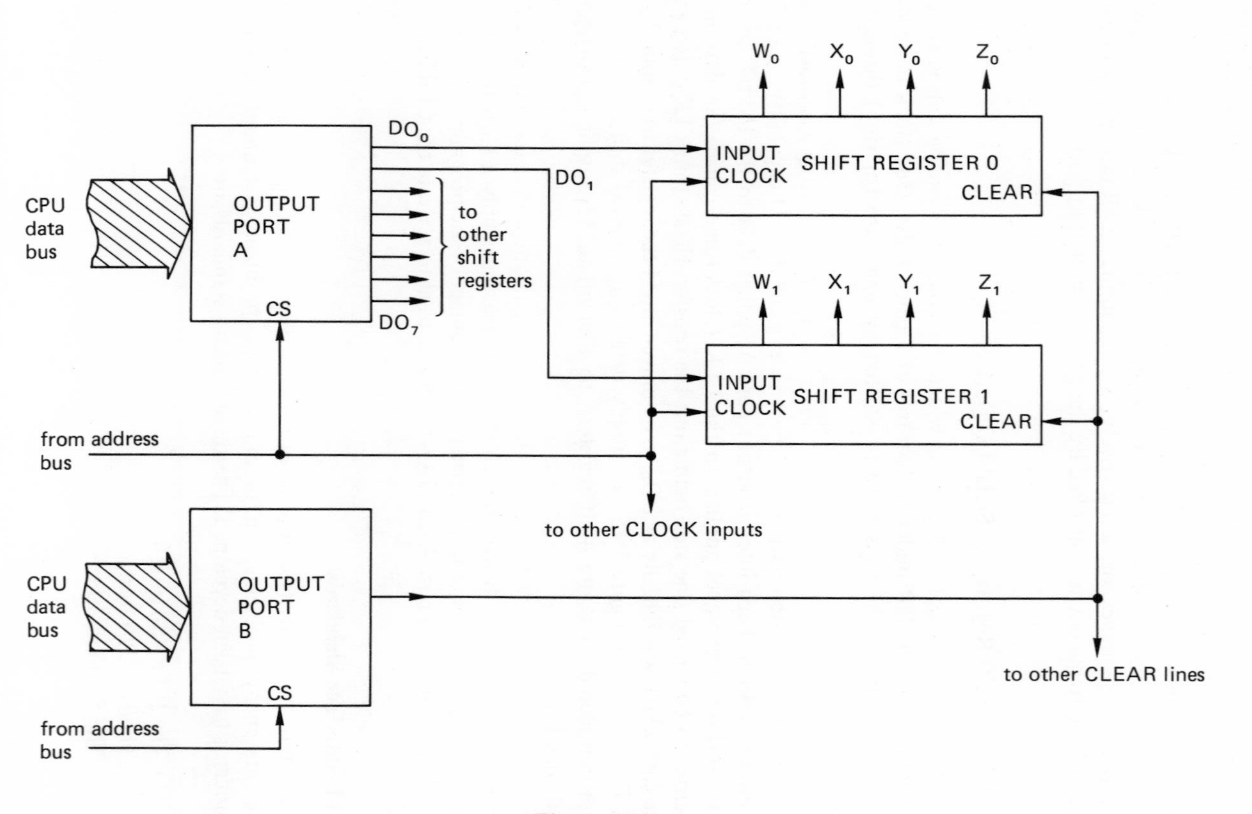

CPU data bus
OUTPUT PORT A
CS
DO0
DO1
DO7
to other shift registers
from address bus
CPU data bus
OUTPUT PORT B
CS
from address bus
W0
X0
Y0
Z0
INPUT
CLOCK
SHIFT REGISTER 0
CLEAR
W1
X1
Y1
Z1
INPUT
CLOCK
SHIFT REGISTER 1
CLEAR
to other CLOCK inputs
to other CLEAR lines

Figure 7.15

National Standards Institute (ANSI). A block diagram showing the general outline of the IEEE-488 interface bus system is shown in figure 7.16. The bus consists of sixteen lines, eight of which are the conventional data bus lines, three are handshake I/O lines, and the remaining five are management lines or controller lines.

Three types of device may be connected, namely **talkers, listeners** and a **controller**. A talker instrument can only transmit data when addressed; a timer or a counter is an example of this type. A listener instrument can only receive information; a printer is an example. A talker/listener instrument can receive data or instructions and later can transmit data; examples include a programmable d.v.m. and an analog-to-digital converter. The controller can, for example, address a specific talker and connect it to a specified listener whilst disabling other listeners.

Other systems which are very similar (or identical to) the IEEE-488 interface are the GPIB (General-Purpose Interface Bus) and the HP-IB (Hewlett-Packard Interface Bus).

The RS-232 interface

The RS-232 standard was devised as an interface between data communications equipment (for example, a modem) and data terminal equipment such as a CRT terminal. Data is transmitted serially, a typical format being that shown in figure 7.13.

Although a 25-pin connector is normally used to connect a device to the RS-232 interface, only three of the connections are required to connect, say, a teletype or CRT terminal to a computer. The three connections are needed for transmitting serial data, receiving serial data, and for the signal return wire (ground).

The RS-232 system is widely used as an interface between computing systems and devices operating at low-to-medium data rates (included in these are teletypes and CRT terminals). Other standards such as the RS-422 and the RS-423 have been devised to deal with the data rates up to 10 megabits per second and 0.1 megabits per second, respectively.

The S-100 interface

The S-100 interface is realised in the form of a 100 contact connector which is mounted on a computer 'mother board', allowing many modules to be connected to the microcomputer system. The S-100 bus connections are grouped into four categories as follows:

(1) Power supplies
(2) Address lines
(3) Data lines
(4) Clock and control lines

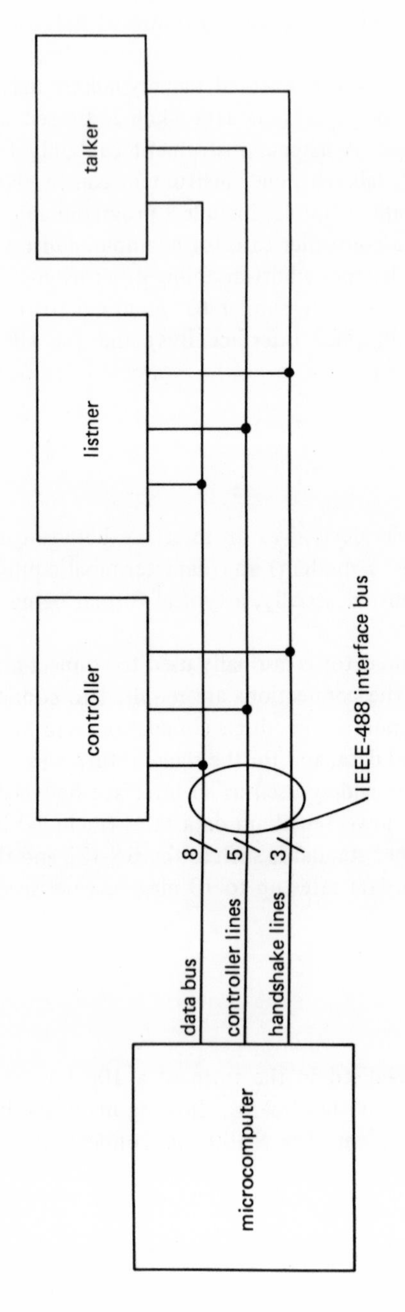
talker
listner
controller
microcomputer
data bus
controller lines
handshake lines
8
5
3
IEEE-488 interface bus

Figure 7.16

PROBLEMS

7.1 Design two logic circuits which fully decode address lines A_7–A_0 to provide (a) active high and (b) active low chip select signals at address 27_{16}.

7.2 Draw a block diagram showing how six 1-out-of-8 decoders can be used to address 48_{10} I/O ports using isolated I/O addressing. Show which address lines are used in association with the decoders, and also indicate how each address is deduced from the address lines used.

7.3 Draw a memory map for a system similar to that in figure 7.4, but in which address line A_9 is logically inverted before being applied to G2, and A_{10} is applied directly to G2.

7.4 Draw a memory map for a system similar to that in figure 7.4, but in which address line A_9 is replaced by A_8, and A_{10} is replaced by A_{11}.

7.5 Describe a typical programmable I/O port containing a small amount of RAM, and outline a typical application of the port.

7.6 Discuss the relative merits of serial data transmission using (a) an on-chip I/O port and (b) a UART.

7.7 If the bit-time of a serial data signal is 0.417 ms, what is the baud rate? Calculate also the maximum character transmission rate if each character has eleven bit-times.

8 Subroutines and Stacks

8.1 The Need for Subroutines

Many programs call for an operation to be repeated several times. For example, every traffic light system needs to introduce a time delay between successive light

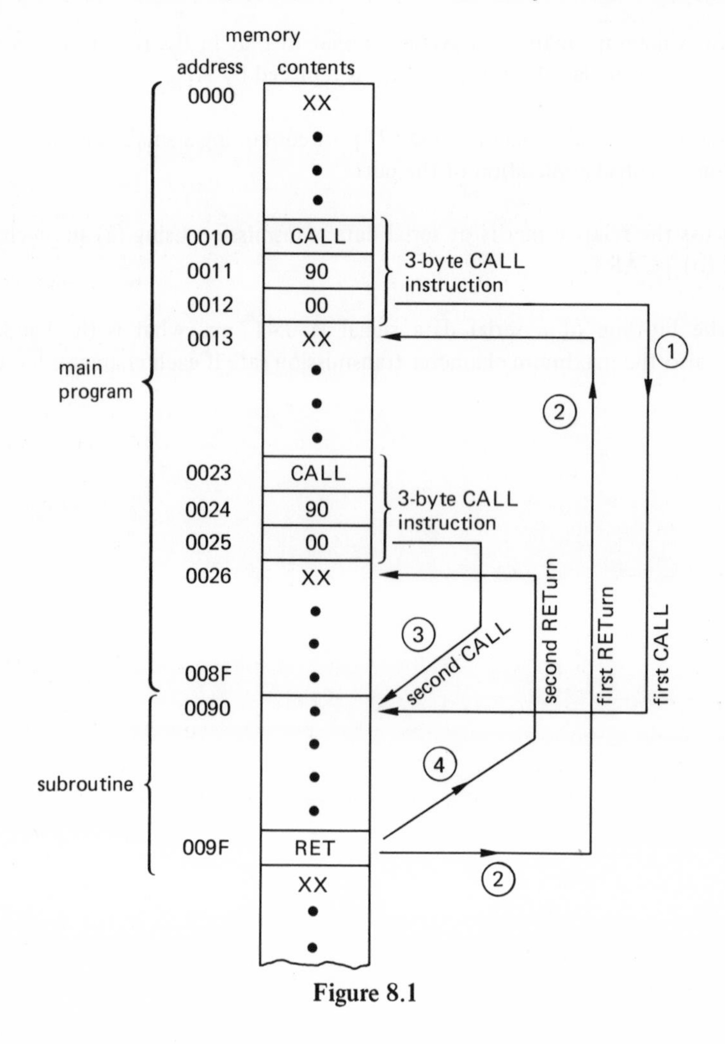

Figure 8.1

display patterns. If the time delay program is written into the main program each time it is required, even in a simple system, it will take up quite a large amount of the available memory space. Clearly, if the time delay program can be written once only, and 'called for' by the main program at appropriate intervals, then not only is the programming effort reduced but also the overall amount of memory space needed for the program is reduced.

The time delay sequence mentioned above can be written in the form of a **subroutine**, that is it is a subprogram not included in the main path of the program. The subroutine is entered by means of a CALL instruction; the return to the main program at the end of the subroutine is made by a RETurn instruction. The basis of the CALL and RETurn sequence is illustrated in figure 8.1. The main program commences at the hex address 0000; at address 0010 the subroutine is CALLed for for the first time. The second byte of the CALL instruction contains the low byte of the address of the subroutine, while the third byte of the instruction contains the high byte of the subroutine address. The CALL instruction causes the microprocessor to transfer control to address 0090 (path ① in figure 8.1), when it commences executing the instructions in the subroutine.

The final instruction in the subroutine is a RETurn instruction at address 009F. This instruction results in program control being returned to the main program at address 0013 via path ② in figure 8.1 (the reason why program control is transferred to this address is discussed in more detail later in the chapter). The above procedure is repeated later in the main program when control is transferred to the subroutine via path ③ in figure 8.1, a RETurn being made to address 0026 via path ④.

Important note: Not all microprocessors have CALL and RETurn instructions in their instruction set (the National Semiconductors INS 8060 CPU does not have them). These microprocessors use one of a number of 'pointer registers' to store the initial address of the subroutine, and also to store the 'return' address when the subroutine has been executed. Such a CPU has special programming requirements when handling a return from a subroutine (and also a return from an interrupt - see chapter 9).

8.2 The Need for a Stack

A **stack** is a region of memory which acts as a temporary store for the contents of the CPU registers and also for the return address for subroutines.

To illustrate the need for a stack, consider the subroutine CALL and RETurn process illustrated in figure 8.1. When jump ① is made from the main program to the subroutine, the microprocessor must 'remember' that the return address is 0013. A common method of 'saving' the return address is to store it in the 'stack'; when the CPU encounters a RETurn instruction at the end of the subroutine, it fetches the return address from the stack and inserts it into the program counter. In this way the CPU ensures that a return is made to the correct point in the program.

When a jump is made to a subroutine, the subroutine may have to use a number of registers which are already storing vital data. This data can be put into temporary storage on the stack by means of a **PUSH** instruction; when the CPU encounters a PUSH instruction, it PUSHes the contents of the appropriate register down on to the stack. The data can later be retrieved by means of a **POP** instruction; when this instruction appears in the program, the data is said to be POPped up from the stack into the appropriate register.

Some CPUs have a stack of limited size, whilst others do not; the latter is clearly advantageous. Additionally, certain CPUs have more powerful stack instructions than others. For example, in the 8080 and similar CPUs it is the programmers responsibility not only to ensure that the register contents are PUSHed on to the stack when entering the subroutine, but also that they are POPped from the stack in the correct sequence when leaving the subroutine. A simple programming error can result in data stored in two registers being interchanged during a PUSH and POP sequence. Certain microprocessors have a single instruction which PUSHes the contents of all the registers on to the stack, and another which POPs them back again. This leads to faster and more reliable handling of subroutine jumps.

8.3 Stack Organisation

Stacks are organised as **Last-in, First-Out (LIFO)** stores. That is to say, the last byte of data stored on the stack is the first to be removed. A simple analogy of a LIFO store is a spring-loaded plate dispenser in a cafeteria. As one plate is loaded on to it the other place at the top of the 'stack' disappears from view; the last plate put on the stack is the first to be removed. However, the analogy has its limitations, since the stack in a microprocessor store does not 'move'.

The initial address of the stack depends on the way in which the microcomputer is organised. In 8080 and similar types of CPU, the initial address is stored under program control in a register known as the **stack pointer** register. In general, a

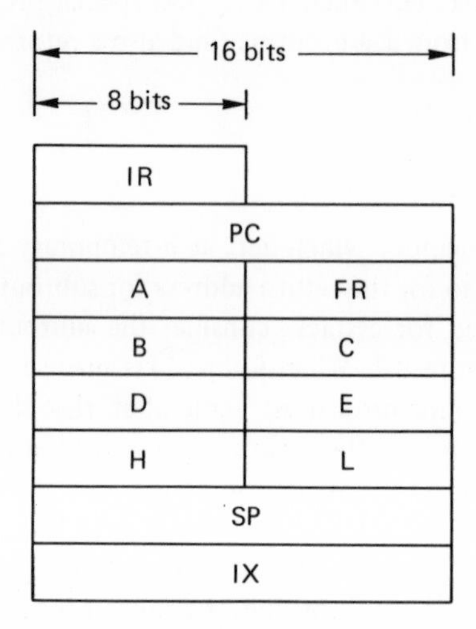

Figure 8.2

pointer register (the stack pointer being one of these) is a register containing an address which 'points' at a particular location in the memory. The function of the stack pointer is to indicate the address of the current 'top' of the stack; consequently, the address stored in the stack pointer changes with the 'height' of the stack.

When it is known that a subroutine is to be used, the programmer must ensure that the stack pointer is loaded (initialised) at an early point in the program with the initial address of the stack. In so doing, the programmer must ensure that he has allocated sufficient memory space for the stack.

The reader has now reached a point in the book where he can look realistically at the organisation of the registers in a typical CPU. These are illustrated in figure 8.2 in which

IR = instruction register (8 bit)
PC = program counter (16 bit)
A = accumulator (8 bit)
FR = flag register (8 bit, but only 5 bits are used in the 8080 – see chapter 4)
B, C, D, E, H, L = auxiliary registers (each 8 bits)
SP = stack pointer (16 bit)
IX = index register (16 bit)

The reader should note that the 8080 family of CPUs does not have an index register (IX), but one has been included in the CPU in this book so that a more varied range of addressing modes can be used.

In the following, a simple program is considered which calls for a subroutine during the execution of the main program (for the convenience of readers, the same addresses are used as in figure 8.1). Prior to discussing the detailed operation of the program, a number of LXI instructions (Load register pair Immediate) will be introduced, one of which is used in the subroutine described below. These instructions are outlined in table 8.1

Table 8.1

Hex machine code	Mnemonic	The instruction causes the CPU to store the contents of the following 16-bit register pair in the selected address
01	LXI B	B and C
11	LXI D	D and E
21	LXI H	H and L
31	LXI SP	Stack pointer (treated as a 16-bit register pair)

Table 8.2

Address	Machine code	Mnemonic	Comment
0000	31	LXI SP	The stack pointer is initialised to address 0109
0001	09		Low byte of the stack address
0002	01		High byte of the stack address
0003	.		
.	.		Main program
.	.		
000A	.		
0010	CD	CALL	Call subroutine
0011	90		Low byte of subroutine address
0012	00		High byte of subroutine address
0013	.		Main program
.	.		
.	.		
0090	XX		Start address of subroutine
.	.		Subroutine
.	.		
009F	C9	RET	Return to main program
.	.		
.	.		
0107	13		Low byte of first return address
0108	00		High byte of first return address
0109	XX		Initial address of stack pointer

The program together with the subroutine and stack is shown in table 8.2. The program commences with a 3-byte LXI SP instruction, which initialises the address of the stack pointer to 0109_{16} (this address is stored in the second and third bytes of the LXI SP instruction). The CPU then steps sequentially through the main program until it reaches the subroutine CALL instruction at address 0010_{16}. This instruction causes program control to be transferred to location 0090_{16}. However, the control unit of the CPU recognises the CALL instruction as a 3-byte instruction, which results in the program counter being incremented to 0013. The CALL instruction carries out three operations as follows

(1) It decrements (reduces by unity) the address in the stack pointer and also transfers the high byte of the address in the program counter to the stack (the stack pointer points now at location 0108).

(2) It decrements the address in the stack pointer once more and transfers the low byte of the address in the program counter to the stack (the stack pointer now points at 0107).

(3) It transfers the subroutine start address to the program counter.

On the completion of the CALL instruction, the contents of the various memory locations are shown in table 8.2. The reader will note that the content of the initial stack address (that is, 0109) is not meaningful data, since the CPU always decrements the stack pointer before writing data into the stack. During the time that the CPU is processing the subroutine, the stack pointer 'points' at address 0107.

On completion of the subroutine, the CPU executes a RETurn instruction which carries out two operations.

(1) It transfers the 'top' byte in the stack (at address 0107) to the low byte of the program counter and increments (increases by unity) the address of the stack pointer (so that it 'points' at address 0108).

(2) It transfers the new 'top' byte of the stack (at address 0108) to the high byte of the program counter and increments the address in the stack pointer (so that it points at 0109).

That is, after the execution of the RETurn instruction the stack pointer is incremented by two (from 0107 to 0109), and the program counter stores the correct return address of 0013. Operations 1 and 2 of the RET instruction leave the contents of locations 0108 and 0107 unchanged at 00_{16} and 13_{16}. The data in these locations is replaced with new data when the next CALL instruction is executed.

The CPU described in this book has a 16-bit stack pointer, which allows it to store the address of the 'top' of the stack. This arrangement not only allows the stack to have any size but also permits it to commence at any address specified by the initial value of the stack pointer. The organisation of stack pointers differs between CPUs; for example, the 6502 CPU (used in APPLE and PET microcomputers) has a stack pointer which is only 8 bits wide. In this case the CPU inserts 01_{16} in the high byte of the stack address, so that addresses 0100 to 01FF are permanently assigned to stack operations; in practice, this size of stack is adequate.

8.4 Use of the Stack for Data and Status Storage

Any subroutine may call for the use of one or more of the registers in the CPU. If these registers already store data determined from earlier operations in the program which must be retained for later use, then it must be 'saved' when a subroutine CALL is made; if the programmer does not save the data, then it will be lost when new data is written into the registers during the execution of the subroutine. Consequently, it is the programmers responsibility to ensure that not only are the contents of the registers saved at the commencement of the subroutine operation, but also that the data is returned to its original register on the completion of the subroutine. A line diagram showing the general organisation of a subroutine is shown in figure 8.3.

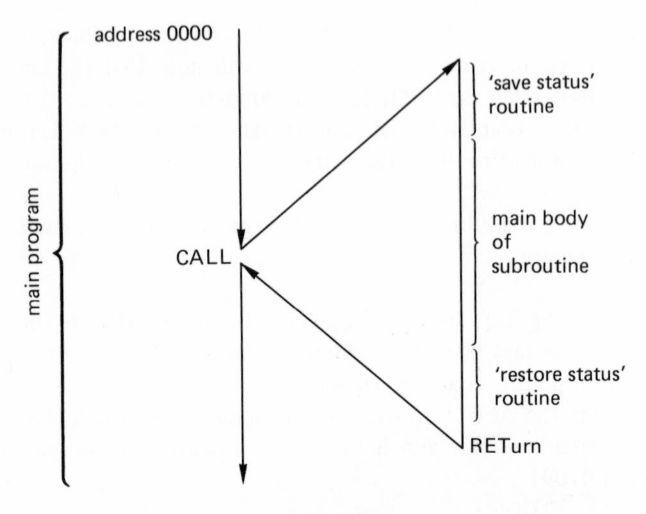

Figure 8.3

The instructions used during the 'save' and 'restore' sections of the subroutine depend on the microprocessor. In this case we use the 8080 family PUSH and POP instructions, which are listed in table 8.3; the reader will recall from earlier work that PSW is the program status word, comprising the flag register and the accumulator.

A subroutine which firstly saves the contents of each user-accessible register and later restores the contents is given in table 8.4; this subroutine uses the same

Table 8.3

Machine code (hex)	Mnemonic	Comment
C1	POP B	POP register pair B,C from stack
C5	PUSH B	PUSH register pair B,C on to stack
D1	POP D	POP register pair D,E from stack
D5	PUSH D	PUSH register pair D,E on to stack
E1	POP H	POP register pair H,L from stack
E5	PUSH H	PUSH register pair H,L on to stack
F1	POP PSW	POP PSW from stack
F5	PUSH PSW	PUSH PSW on to stack

Table 8.4

Address	Machine code	Mnemonic	Comment
0090	C5	PUSH B	Save data routine
0091	D5	PUSH D	
0092	E5	PUSH H	
0093	F5	PUSH PSW	
0094			Main body of subroutine
.			
.			
.			
009A			
009B	F1	POP PSW	Restore data routine
009C	E1	POP H	
009D	D1	POP D	
009E	C1	POP B	
009F	C9	RET	Return to main program

range of addresses as those in table 8.2. Using an initial stack address of 0109, on executing a CALL instruction the CPU stores the program counter address in locations 0108 (high byte) and 0107 (low byte), followed by register pair BC, then pair DE, then pair HL and, finally, the PSW. Each PUSH instruction causes the stack pointer to be decremented by two, so that after PUSH B it points to 0105, after PUSH D it points at 0103, etc. During the execution of the main body of the subroutine, the stack pointer contains address 00FF (see table 8.5). Towards the end of the subroutine (during instructions 009B – 009E), the 'restore data' routine causes the original data stored in the registers to be restored to them in the correct sequence. To do this **the POP sequence must be executed in the reverse order to that of the PUSH sequence.** Each POP instruction causes the stack pointer to be incremented by two, so that after POP PSW the stack pointer points at 0101_{16}, after POP H it points at 0103_{16}, etc. The mechanics of the PUSH D and POP D instructions are illustrated in figure 8.4.

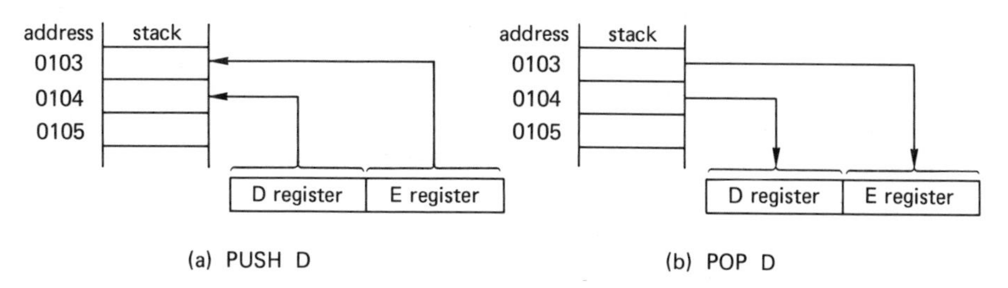

Figure 8.4

Table 8.5

Address	Stack contents	Comment
00FF	flags	Stack pointer address during the main body of the subroutine
0100	accumulator	
0101	L register	
0102	H register	
0103	E register	
0104	D register	
0105	C register	
0106	B register	
0107	PC low byte	
0108	PC high byte	
0109	XX	Initial address of stack pointer

If PUSH and POP instructions are reversed when compared with the above sequence, then the data in the registers will be reversed at the end of the subroutine. For example, if the program includes the instructions

```
PUSH B
PUSH D
  .
  .
  .
POP B
POP D
```

then the contents of the register pairs BC and DE are interchanged (which may be the required result in certain instances).

8.5 A Time Delay Subroutine

An example of a practical subroutine is described in this section, the example chosen being that of a time delay. It was stated earlier that a time delay is generated simply by making the CPU execute a large number of time-wasting instructions. To generate a known (and accurate) time delay, the programmer must know

(1) The periodic time (the cycle time) of the CPU timing oscillator.
(2) The number of clock cycles needed for each instruction.
(3) The number and type of instructions executed.

The cycle time of the clock generator depends on the CPU; many CPUs use a 2-MHz clock having a periodic time of

$$\frac{1}{\text{clock frequency}} = \frac{1}{2 \times 10^6} = 0.5 \times 10^{-6} \text{ s or } 0.5\ \mu\text{s}$$

The number of clock cycles needed to execute an instruction is specified in the CPU literature, and the following are typical of 8080 and 8085 CPUs.

Instruction	Clock cycles
MVI r	7
DCR r	5
JNZ	10

The MVI r instruction above causes the CPU to MoVe data Immediately into a register (one of the registers A, B, C, D, E, H or L). The DCR r instruction causes the CPU to DeCRement (reduce by unity) the contents of a register, and the JNZ instruction causes program control to Jump on No Zero (that is, when the zero status flag is not set) to another address. The three instructions above are combined to give a 0.2-s time delay by means of the subroutine listed in table 8.6; the corresponding flowchart is shown in figure 8.5.

The subroutine calls for register B to be loaded with zero by means of a 2-byte MVI B,00_{16} instruction. This is followed by a 2-byte MVI C,68_{16} instruction which loads register C with 68_{16} or 104_{10}. Next, the DCR B instruction causes register B to be decremented, so that it stores FF_{16}. The 3-byte JNZ instruction commencing

Table 8.6 A 0.2-second time delay subroutine

Address	Machine code	Mnemonic	Comment
0110	06	MVI B	Load register B with 00_{16}
0111	00	00_{16}	
0112	0E	MVI C	Load register C with 68_{16} (= 104_{10})
0113	68	68_{16}	
0114	05	DCR B	(B) = (B) − 1
0115	C2	JNZ	Jump if (B) not zero to address 0114
0116	14		Low byte of 'jump' address
0117	01		High byte of 'jump' address
0118	0D	DCR C	(C) = (C) − 1
0119	C2	JNZ	Jump if (C) not zero to address 0114
011A	14		Low byte of jump address
011B	01		High byte of jump address
011C	C9	RET	Return to main program

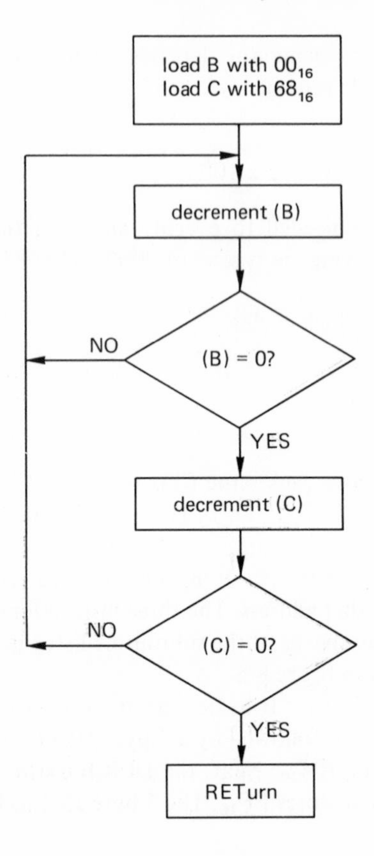

Figure 8.5

at location 0115 results in the CPU transferring control to the DCR B instruction so long as the contents of register B are non-zero; this program loop is therefore executed $(FF + 1)_{16} = 256_{10}$ times before the contents of register B are reduced to zero. From the information given earlier, fifteen clock cycles are needed to execute each pass of the DCR B and JNZ loop. Since our CPU has a 0.5 μs cycle time, the time taken to escape from this loop for the first time is

$$256 \times (15 \times 0.5 \times 10^{-6})\ \text{s} = 1.92\ \text{ms}$$

The CPU then decrements the contents of register C and, so long as the contents of register C are non-zero, it returns to the DCR B instruction again. Since register C originally contained 104_{10}, the total time delay associated with this part of the program is

$$104 \times 1.92\ \text{ms} = 0.1997\ \text{s}$$

Other instructions in the subroutine add a tiny amount of time to this figure and, for all practical purposes, the subroutine can be regarded as generating a 0.2-s time delay. The time delay can be altered by modifying the data stored in location 0113; the delay becomes 0.25 s by storing 82_{16} (130_{10}) in this location, and is 0.1 s if 34_{16} (52_{10}) is stored.

The machine code list in the second column of table 8.6 is known as an **object program**, and consists of instructions in the machine language (although in our case it is in hex rather than pure binary) which can be understood directly by the CPU. To make the program more easily understandable to humans, the instructions are usually written in the first instance in the form of mnemonics (a **symbolic language**) as shown in the third column of table 8.6. The latter program is known as a **source program**. As mentioned earlier in the book, the source program is converted into the object program by means of an assembler program. Assembly language instructions (or statements) are divided into a number of **fields** (a field is a subdivision of a record of some kind; for example a company payroll record will have a number of fields including company reference number, gross pay, tax code, superannuation contribution, etc). The fields of an assembly language instruction are:

(1) Label field
(2) Mnemonic field
(3) Operand or address field
(4) Comment field

A **label is a group** of characters (for example, DELAY 1 to represent a time delay program) which identifies an instruction or item of data or memory location; a label is translated into an address by the assembler. A **mnemonic** is a symbolic code which enables the user to recall the function of the machine code; for example, the MVI mnemonic represents MoVe Immediate. The **operand or address field** contains the data stored in the second (or second and third) byte of the instruction. The **comment field** contains brief notes which clarify the program procedure, but have no effect on the CPU itself since the comments are not translated into machine code by the assembler.

Table 8.7 Source program for a 0.2-second time delay

Label field		Mnemonic field	Operand/address field	Comment field
TIME 1	:	MVI	B,00_{16}	; GET COUNT FOR DELAY
		MVI	C,68_{16}	; GET VALUE FOR PERIOD
DELAY	:	DCR	B	; COUNT = COUNT − 1
		JNZ	DELAY	; CONTINUE UNTIL COUNT = 0
PERIOD	:	DCR	C	; VALUE = VALUE − 1
		JNZ	DELAY	; CONTINUE UNTIL VALUE = 0
		RET		; RETURN TO MAIN PROGRAM

Depending on the program, fields 1, 3 and 4 above could be empty, but the mnemonic field (field 2) must always contain an instruction mnemonic. Moreover, the assembler must have some method of knowing when one field ends and the next commences; to discriminate between fields, assemblers use a special symbol (a **delimiter**) either at the beginning or end of each field. Typical delimiters are commas, spaces, semicolons, colons, slashes and question marks. In the following, a colon is used to separate the label field from the mnemonic field, a space is left between the mnemonic field and the operand/address field, and a semicolon separates the operand/address field from the comment field.

To illustrate the use of a source program, the 0.2-s time delay program in table 8.6 is rewritten in table 8.7 in source program format. The reader will note that addresses are no longer given, but are implied by statements in the label field.

8.6 Nested Subroutines

It is frequently necessary for one or more subroutines (which we will describe as **low-level subroutines**) to be called or used by another subroutine (which we will call a **high-level subroutine**). For example, a 20-s time delay subroutine (a high-level subroutine) can be generated by using the TIME 1 subroutine (the low-level subroutine) in table 8.7 one hundred (64_{16}) times. This is illustrated in the flowchart in figure 8.6. Here a program labelled TIME 2 produces a 20-s time delay by calling for the 0.2-s time delay generated by TIME 1 one hundred times; since the stack deals with the return addresses of a number of subroutines, the programmer must

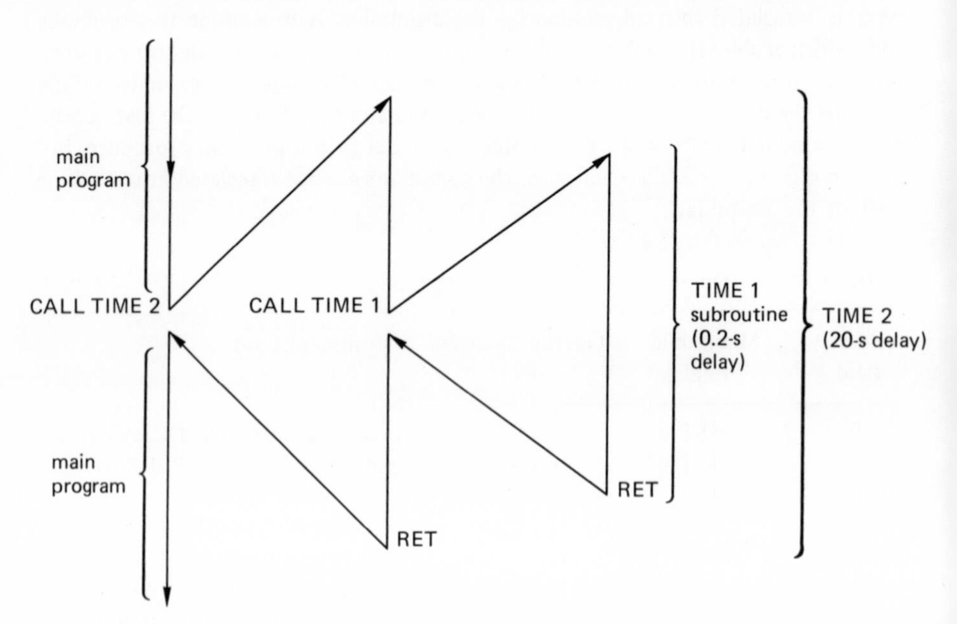

Figure 8.6

Table 8.8 Generating a 20-second time delay using nested subroutines

Label field		Mnemonic field	Operand/address field	Comment field
TIME 2	:	MVI	E,64_{16}	; GET NUMBER FOR AGAIN
AGAIN	:	CALL	TIME 1	; CALL 0.2s DELAY
		DCR	E	; NUMBER = NUMBER - 1
		JNZ	AGAIN	; CONTINUE UNTIL NUMBER = 0
		RET		; RETURN TO MAIN PROGRAM

ensure that he has reserved enough storage space in the stack. A source program for a 20-s time delay is given in table 8.8.

In the case of the subroutine TIME 2, register E is loaded with 64_{16} (100_{10}), and TIME 1 is called for 100_{10} times. On each occasion that TIME 1 is called, the address placed at the top of the stack is the correct return address to the TIME 2 subroutine, that is, it returns to the DCR E (DeCRement register E) instruction. Since the time taken to complete TIME 2 is (100 x 0.2) = 20 s, a return to the main program is made when the content of register E is zero.

The **nesting depth** may have any value, with one subroutine calling a second subroutine, which may in turn call a third subroutine and so on.

Subroutines provide a convenient way of producing manageable sections of programs, and can be regarded as an aid in preserving the programmer's sanity when dealing with a complex problem. However, since subroutines use the stack, they are more difficult to debug than a normal program. The reason for saying this, is that the contents of the stack pointer may change during the operation of the subroutine, and errors in stack usage can be difficult to resolve.

8.7 Subroutine Calls using the RESTART Instruction

The 8080 family of microprocessors has a number of single-byte subroutine call instructions known as RST (ReSTart) instructions. When a RST instruction is executed, it results in the following actions:

(1) The current contents of the program counter are automatically pushed on to the stack, and the stack pointer is decremented by two.

(2) Program control is transferred to the instruction whose address is related to the 'value' of the RST instruction, that is, a program jump is made to a specified address.

That is, a RST instruction is another type of CALL instruction. The RST instruction is written in the source program in the form

RST *n*

where n is an integer in the range zero to seven. The 8-bit word representing the RST instruction is as follows

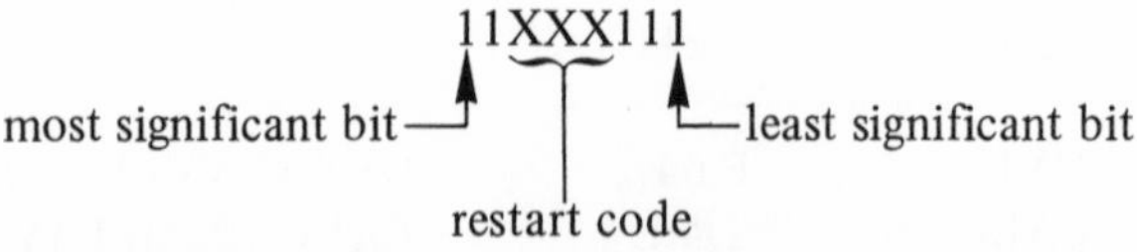

11XXX111

most significant bit — restart code — least significant bit

where each 'X' is either a '0' or a '1'. The decimal value of the binary combination of X's specifies the restart 'number' as shown in table 8.9.

Table 8.9 Restart numbers

Instruction	Bit combination XXX	Hex value of RST instruction (its op-code)
RST 0	000	C7
RST 1	001	CF
RST 2	010	D7
RST 3	011	DF
RST 4	100	E7
RST 5	101	EF
RST 6	110	F7
RST 7	111	FF

The RESTART instruction causes the following binary combination to be transferred into the program counter

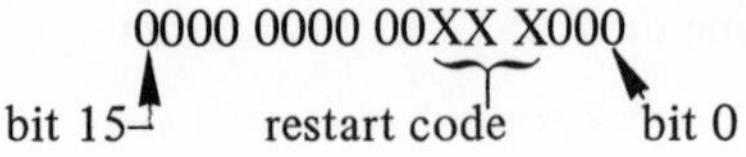

0000 0000 00XX X000

bit 15 — restart code — bit 0

where the group XXX has the value given in table 8.9. Thus a RST 0 instruction results in a jump to the following address

$$0000\ 0000\ 0000\ 0000_2 \quad \text{or} \quad 0000_{16}$$

That is, a RST 0 instruction is equivalent to a hardware RESET signal. A RST 5 instruction results in a jump to address

$$0000\ 0000\ 0010\ 1000_2 \text{ or } 0028_{16}$$

A complete list of RST instructions together with their starting addresses is given in table 8.10. An investigation of table 8.10 shows that the RST instruction causes program control to be transferred to an address which is 8_{10} times the value of the

Table 8.10

Instruction mnemonic	Op-code	Subroutine start address (hex)
RST 0	C7	0000
RST 1	CF	0008
RST 2	D7	0010
RST 3	DF	0018
RST 4	E7	0020
RST 5	EF	0028
RST 6	F7	0030
RST 7	FF	0038

RESTART code. Thus RST 5 results in control being transferred to address $(5 \times 8)_{10} = 40_{10} = 28_{16}$. A limitation of the use of the RST instruction is that, in general, only eight instructions can be inserted between RST numbers. However, the RST instruction is widely used in the 8080 family of CPUs for providing program interrupts (see chapter 9).

8.8 Conditional CALL and RETurn Instructions

The CALL instruction described earlier results in the CALLed subroutine being executed unconditionally. However, there are circumstances in which a subroutine must only be called under specified conditions, and ignored under other conditions. All microprocessors have some form of **conditional CALL** instructions which, in the case of the CPU considered here, are 3-byte instructions. The first byte defines the conditions under which the CALL is made, the remaining two bytes give the starting address of the subroutine. The conditional CALL instructions are listed in table 8.11.

Table 8.11

Instruction mnemonic	Op-code	Condition for CALL
CNZ	C4	No Zero (zero flag = 0)
CZ	CC	Zero (zero flag = 1)
CNC	D4	No Carry (carry flag = 0)
CC	DC	Carry (carry flag = 1)
CPO	E4	Parity Odd (parity flag = 0)
CPE	EC	Parity Even (parity flag = 1)
CP	F4	Plus (sign flag = 0)
CM	FC	Minus (sign flag = 1)

Table 8.12

Instruction mnemonic	Op-code	Condition for RETURN
RNZ	C0	No Zero (zero flag = 0)
RZ	C8	Zero (zero flag = 1)
RNC	D0	No Carry (carry flag = 0)
RC	D8	Carry (carry flag = 1)
RPO	E0	Parity Odd (parity flag = 0)
RPE	E8	Parity Even (parity flag = 1)
RP	F0	Plus (sign flag = 0)
RM	F8	Minus (sign flag = 1)

Any subroutine in our CPU may be terminated by one of a number of 1-byte conditional return instructions. The eight conditional return instructions used by the 8080 family are listed in table 8.12.

PROBLEMS

8.1 Explain how data is loaded into and removed from a microcomputer stack.

8.2 A microcomputer system has a total of 8K words of RAM. If the first available memory location is at address 0000_{16}, what address must be loaded into the stack pointer to ensure that every location in the stack can be used.

8.3 If the microprocessor in problem 8.2 is modified so that it has a total of 28K words or RAM, what address must be loaded into the stack pointer to ensure that the first storage byte in the stack is located at the very top of the memory space?

8.4 Write a POP and PUSH sequence which will cause the contents of the register pair BC to be interchanged with HL.

8.5 Write a time delay subroutine which generates a time delay of 0.15 s.

8.6 Write a nested subroutine which generates a time delay of 150 s; use the subroutine written for problem 8.5 as the basic timing sequence.

8.7 Show how a RESTART instruction can be used to call the subroutine in problem 8.6.

9 *Interrupts and Polling*

9.1 The meaning of Polling and Interrupts

Many microcomputers have a large number of input devices connected to them. Since the CPU can only receive data from one input device at a time, the programmer may be in a dilemma as to which device to take data from at a particular instant in time. Take, for example, a steam boiler control system in which the steam pressure is monitored by a sensor. If the pressure suddenly rises to an unacceptably high value, the CPU must take rapid action to reduce the pressure. However, since the CPU may have to read data from a very large number of sensors, the question arises as to how the computer can be organised to detect quickly which sensor indicates a dangerous condition.

A somewhat similar position arises in general-purpose microcomputers which have a range of devices connected to them including switches, keyboards, printers, cassettes, modems, etc. Some of the devices operate at low speed (such as a manually operated switch, which is either ON or it is OFF for a minimum period of a fraction of a second or so), while others operate at higher speeds (a modem is an example of this kind). If data is received simultaneously from a switch and a modem, the programmer must ensure that the data from the modem has higher priority than that from the switch, since the data from the latter is available for a longer time. The dilemma can be resolved either by polling or by an interrupt.

Polling is used to describe a method of periodic interrogation of each device in turn to determine whether it requires servicing. In many cases, such a search reveals that none of the devices requires the attention of the CPU. Consequently, the CPU may spend much of its time searching for a peripheral which is ready for use. The primary disadvantage of polling over interrupts (see below) as a method of evaluating when a peripheral is ready for use is that it is expensive on software and on operating time. Its advantage is that its hardware requirements are low when compared with an interrupt-driven system.

An **interrupt** is a means of producing a subroutine call using hardware external to the CPU, for example, the interrupt call may be generated by an I/O port. The procedure involved is, with certain differences, similar to the CALL instruction described in chapter 8. An interrupt can be described as an interruption in the normal running of a program, and is caused by an external device which results in the program control being transferred to a subroutine (known as the **interrupt routine**). The interrupt is made in such a way that an orderly return is finally made to the point in the main program where the interruption occurred.

To illustrate the use of an interrupt, consider the steam boiler control scheme mentioned earlier. When the steam pressure monitor detects excessive pressure, it causes a signal to be applied to the 'interrupt' pin of the CPU. The control unit of the CPU allows the microcomputer to complete the instruction it is at present executing, after which it suspends operation on the main program for the time being. Program control is then handed to a subroutine specified by the interrupt signal; in the case considered, the subroutine may cause an alarm to be sounded and, quite possibly, it may result in some action being taken to reduce the steam pressure. On completing the interrupt routine the CPU executes a RETurn instruction, allowing program control to be returned to the main program at the point where operations were suspended.

Interrupts can service I/O devices very quickly and are inexpensive on software, but they require more hardware than polling methods. As with subroutines, interrupt-driven programs are relatively difficult to test and debug.

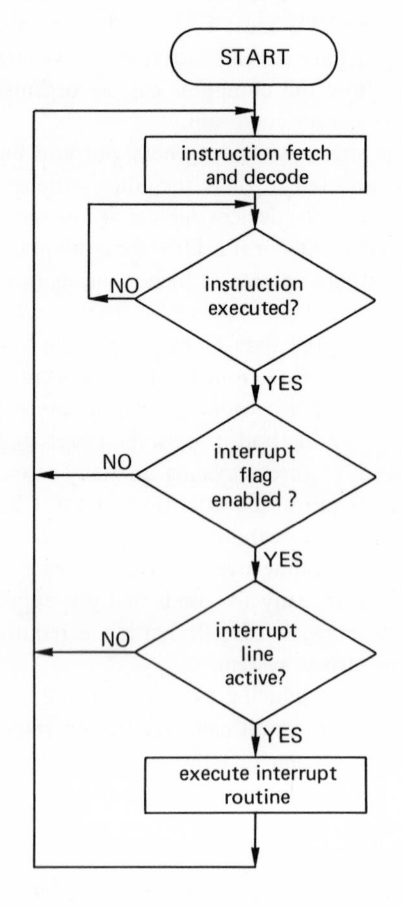

Figure 9.1

9.2 Basic Features of an Interrupt System

The basic operations within a CPU using a program which may contain an interrupt are shown in figure 9.1. The control unit interrogates the state of the CPU during each instruction cycle and, on the completion of the instruction execution cycle, it checks the state of the **interrupt flag**. This is a flag or flip-flop in the CPU which can either be enabled or disabled* by means of a software instruction.

If the interrupt flag is not enabled, then the CPU can fetch and execute the next instruction in the sequence. If the interrupt flag is set, the CPU proceeds to interrogate the state of the interrupt pin on the CPU chip (some CPUs have more than one interrupt pin, but we confine our attention here to a CPU with a single interrupt line). If the interrupt line is not active, then the CPU fetches and executes the next instruction in the program.

Should the interrupt line be active, control is transferred from the main program to the interrupt sequence. On completing the interrupt program, the CPU returns to the main program. As with other subroutine operations, the stack is used to store the return address to the main program.

The way in which the CPU responds to an interrupt signal depends on the design of the CPU. One popular method is to cause the CPU to generate an **interrupt acknowledge** signal ($\overline{\text{INTA}}$ - active low), which may either be available on the CPU chip itself or be obtained from a control bus decoding circuit. The interrupt acknowledge signal is used to activate a device select pin on the I/O port which requests the interrupt service.

9.3 Types of Interrupt

Broadly speaking, interrupts may be classified as being either non-maskable or maskable.

A **non-maskable interrupt** is one which the programmer cannot disable or remove under program control; when the non-maskable interrupt pin is activated, it causes the program to branch to a fixed address in memory (which is address 0024 in the Intel 8085 and address 0066 in the Zilog Z80). A non-maskable interrupt is used in association with a subroutine having the highest possible priority; an example of this type is a 'power failure' interrupt, which must have priority over all other operations. In the few milliseconds between a power failure being detected (that is, a low voltage) and the time when the supply voltage is too low for the CPU to function properly, the program must branch to the power supply failure interrupt routine which 'saves' vital data in an area of the memory which has its own battery power supply.

A **maskable interrupt** is one which the programmer can disable or mask out under program control. The programmer may, for example, wish to disable the interrupts while the CPU is carrying out certain calculations or is executing a timing

*Certain CPUs have non-maskable interrupts (described in section 9.3) which cannot be disabled by means of a software instruction.

sequence. Resetting the CPU to address 0000_{16} automatically disables maskable interrupts, and the programmer must insert an EI (Enable Interrupt) instruction at an early point in the program, otherwise the CPU will ignore signals applied to the interrupt line.

However, a feature of microcomputers is that the CPU does not enable the interrupts immediately it has read the EI instruction from the program; the interrupt enabling process is delayed until the computer has executed one more instruction as described below. Many interrupt routines finish with the following sequence

```
EI     ;ENABLE INTERRUPTS
RET    ;RETURN TO INTERRUPTED PROGRAM
```

If a logic '1' is applied to the interrupt pin of the CPU at the instant that the EI instruction is executed, the signal is ignored until the RET instruction has been executed. This allows the CPU to make an orderly return to the main program before it interrogates the state of the interrupt line. The interrupt input can be disabled at any time by including a DI (Disable Interrupts) instruction in the program.

When control is transferred to an interrupt routine, the CPU jumps either to a fixed address or to a vectored address. A **fixed address interrupt** causes a branch to a specific location in the memory; an example of this type is the non-maskable interrupt mentioned in the first paragraph of this section. A **vectored interrupt** causes program control to be vectored towards or directed towards one of a number of addresses (an RST instruction is an example of this type - see section 9.4).

The number of interrupt pins on the CPU chip depends on its design. Some have a single interrupt line and, for simplicity, the CPU considered in this book has a single input line. Others have several interrupt lines, each having a different order of priority; this allows a more important interrupt routine to interrupt the execution of a lower-order interrupt sequence; that is, interrupts can be 'nested'. However, by the use of additional hardware (such as priority encoders), a CPU with a single interrupt line can provide differing orders of interrupt priority (see section 9.8).

9.4 Vectored Interrupts

In the CPU considered in this book, the RST (RESTART) instruction is used to produce a vectored interrupt. This instruction was described in section 8.7 in connection with a subroutine call instruction. However, in the case of an interrupt, the RST *n* instruction is generated by means of a port external to the CPU as follows.

When the interrupt pin of the CPU is activated by a signal from a device which requires immediate service, the CPU sets aside the program it is currently handling, and outputs an address on the address bus which selects the interrupting port. At the same time it outputs an $\overline{\text{INTA}}$ signal on the control bus which acknowledges receipt of the interrupt signal (circuit details are described later). When this occurs, the interrupting port is allowed access to the data bus, at which point it causes the appropriate RESTART instruction to be **jammed** into the CPU. The net result of the above is that program control is transferred to the start address of a subroutine appropriate to the RESTART instruction used (for details see table 8.10). Thus if the instruction jammed into the CPU corresponds to RST 3, the first instruction in the interrupt sequence commences at address 0018_{16}.

9.5 Interrupt Methods

A number of popular maskable interrupt techniques are shown in figure 9.2. Diagram (a) shows a simple single-line interrupt system with only one device on the system which can produce an interrupt signal. An example of this kind is a motor control scheme which has a single stop button. When the STOP button is pressed it activates the interrupt line of the CPU, resulting in some appropriate action determined by the interrupt routine.

Diagram (b) illustrates a system which deals with several interrupting devices, yet has only one interrupt line on the CPU. The interrupt signals from the devices are ORed together to produce an interrupt signal. The interrupt routine must cause the CPU to scan each of the interrupting devices to determine which one produced the interrupt signal. Multilevel (priority) interrupts can be provided by means of additional hardware.

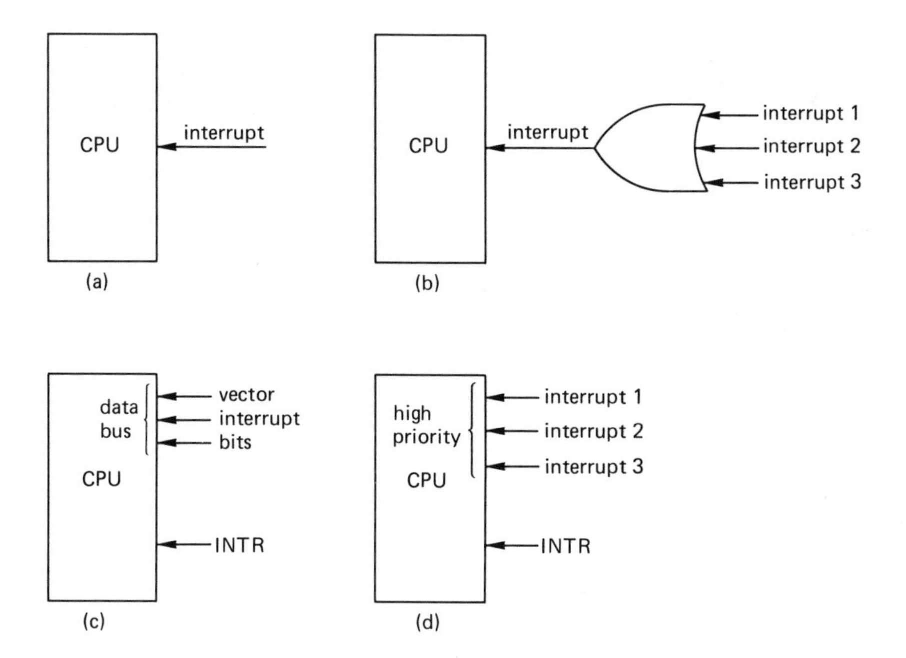

Figure 9.2

The basis of a vectored interrupt is shown in figure 9.2c. When the interrupt line is activated the interrupting port is given access to the data bus, and the byte it jams on the data bus causes a branch to the appropriate RESTART address. If additional hardware is incorporated, multilevel (priority) interrupts can be provided.

A CPU with an on-chip multilevel interrupt facility is shown in figure 9.2d, each interrupt line being 'tied' to a separate interrupting device. In addition to the three multilevel interrupts (1, 2 and 3), the CPU also has a single-line interrupt marked INTR, which permits the use of a vectored interrupt facility. Interrupts 1, 2 and 3 have a higher priority than the INTR line, and take precedence over it should one of them be activated.

In addition to the maskable interrupt facilities in figure 9.2, the CPU may also have a non-maskable interrupt line (see section 9.3).

9.6 Vectored Interrupt Hardware

Figure 9.3 shows a simple interrupt port which uses a number of readily available gates. The system can deal with up to four interrupt signals (INT 7, 6, 5 and 3, which generate the RESTART signals RST 7, RST 6, RST 5 and RST 3, respectively).

Should any interrupt line be driven high, then a logic '1' is applied to the INTerrupt line of the CPU. This results in the CPU suspending operations on the main program and, simultaneously, it forces the $\overline{\text{INTA}}$ line low. The latter signal enables the inverting buffers G2–G4, which jam the appropriate RST signal on to the data bus as follows.

Suppose for the moment that INTerrupt device 6 requires servicing. In this case the signal on line INT 6 is high, while the signals on lines INT 5 and INT 3 remain low. When the inverting buffers are enabled, it results in the following conditions being forced on to the data bus lines

$$D_5 = 1, D_4 = 1, D_3 = 0$$

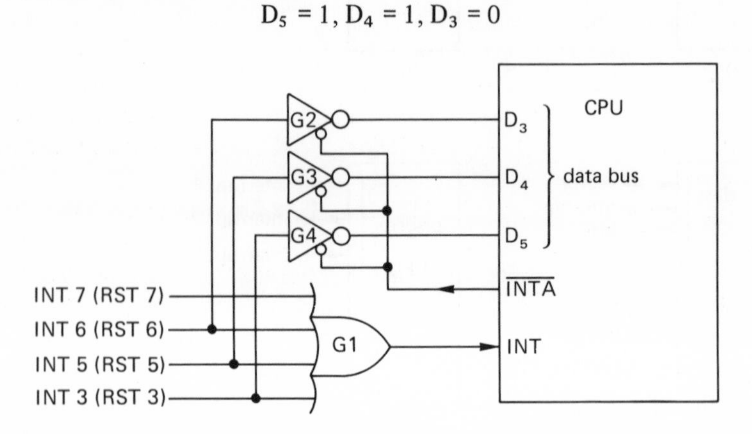

Figure 9.3

At the same time, other I/O devices connected to the data bus lines are disabled so that the remaining five data bus lines (D_7, D_6, D_2, D_1, D_0) are left 'floating'. The reader will recall from earlier work that a 'floating' line appears to the CPU as though it had a logic '1' on it. Thus with INT 6 active, the signals on the data bus lines are

$$\underbrace{11110111_2}_{\text{RST 6}}$$

which is equivalent to the hex word F7 (see also table 8.9). The CPU converts this into a RST 6 instruction, and program control is transferred to address 0030_{16} (see table 8.10). The reader would find it an interesting exercise to verify the operation for INT 5 and INT 3.

When INTerrupt 7 in figure 9.3 is forced high, the logic signals on INT 6, INT 5 and INT 3 are low. Consequently, when the inverting buffers are enabled, the hex byte FF is forced on to the data bus. The CPU interprets this as an RST 7 instruction, and control is transferred to address 0038_{16}.

A simple vectored interrupt system is shown in figure 9.4; this circuit uses I/O ports of the type described in chapter 3, both ports being connected to operate as input ports. Port A is described as an **interrupt instruction port** which applies the RESTART instruction to the data bus. Port B simply acts as an input port which connects the device requesting service to the data bus. A typical interrupting device may be a keyboard; when the operator strikes a key, the keyboard generates an 8-bit word corresponding to the key operated and places it on the input lines to port B. At the same time it also generates a pulse at its output terminal marked INTR; this pulse is applied to the INTerrupt line of the CPU. The CPU acknowledges the interrupt signal by forcing its $\overline{\text{INTA}}$ line low; in turn, this signal enables port A to apply a RESTART code (RST 3 in this case) on to the data bus. The CPU accepts the RESTART signal as an instruction, and transfers program control to the subroutine which commences at address 0018_{16}.

The interrupt routine results in an INPUT PORT READ from port B (port B is an accumulator I/O addressed device at address 01_{16} – see the connections to DS2 on port B [also refer to figure 5.1]). Thus the data from the interrupting device is transmitted via port B to the accumulator of the CPU. A program which carries out the above sequence is outlined in table 9.1.

The first instruction in table 9.1 initialises the stack pointer address at 0100_{16}; the programmer must ensure that this address leaves sufficient memory space for the stack. The EI instruction results in the interrupt flag being enabled, but not until one more instruction is executed; this instruction being the JUMP to the main program. The 3-byte JuMP instruction results in a jump around the range of addresses which are reserved for the RST instructions (see also table 8.10).

When the CPU interrupt line is activated by a signal from an interrupting device (see figure 9.4), program control is transferred to address 0018_{16}. A feature of the interrupt sequence is that the CPU automatically disables the interrupt flag before

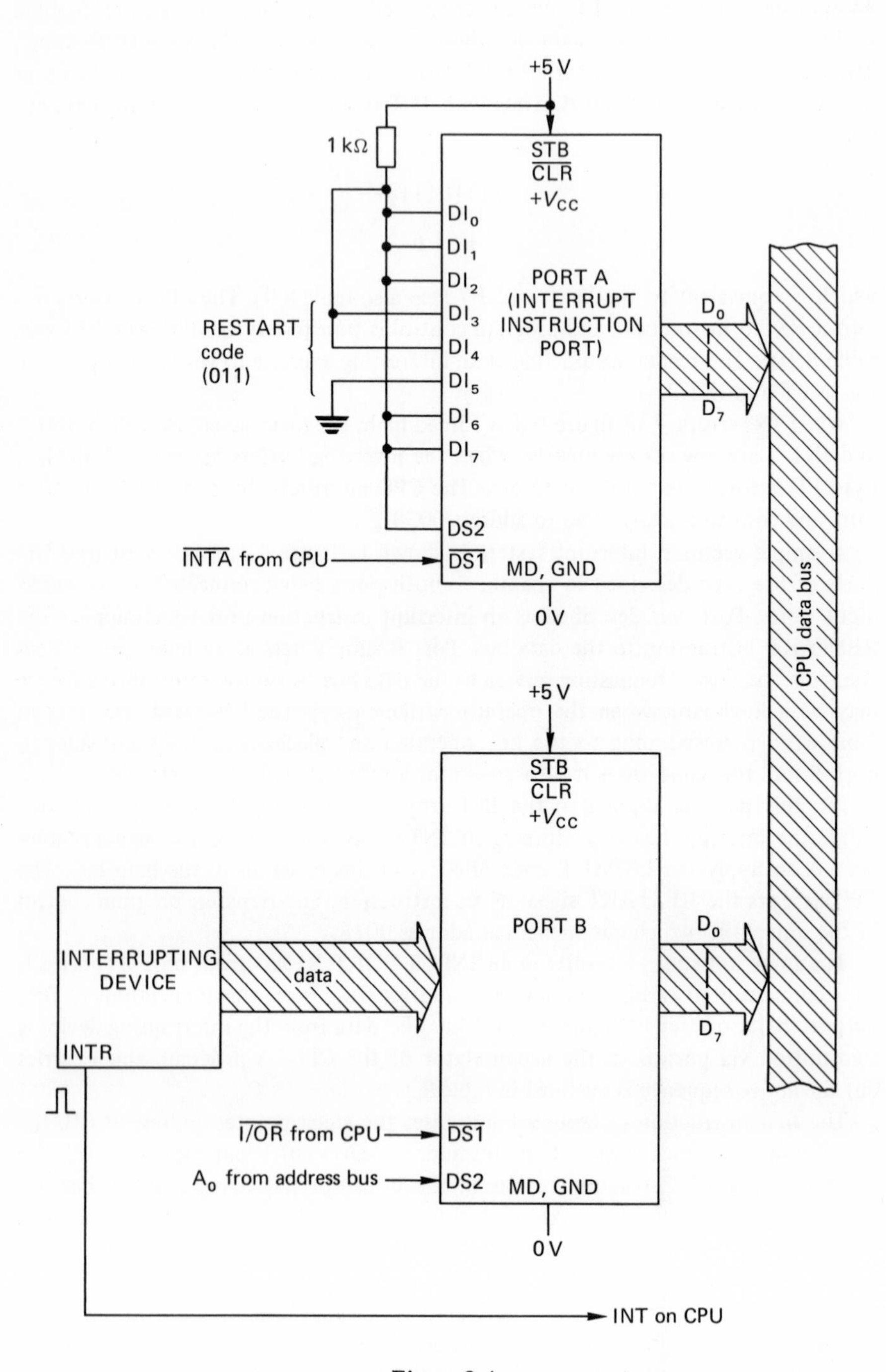
+5 V
1 kΩ
STB
CLR
+V_{CC}
DI_0
DI_1
DI_2
DI_3
DI_4
DI_5
DI_6
DI_7
PORT A
(INTERRUPT
INSTRUCTION
PORT)
RESTART
code
(011)
D_0
D_7
DS2
INTA from CPU
DS1
MD, GND
0 V
CPU data bus
+5 V
STB
CLR
+V_{CC}
INTERRUPTING
DEVICE
data
PORT B
D_0
D_7
INTR
I/OR from CPU
DS1
A_0 from address bus
DS2
MD, GND
0 V
INT on CPU

Figure 9.4

Table 9.1

Address		Instruction		Comment
0000	START:	LXI	SP,0100_{16}	; INITIALISE STACK POINTER ADDRESS
0003		EI		; ENABLE INTERRUPT FLAG
0004		JMP	MAIN	; JUMP TO MAIN PROGRAM
.				
.				
0018	interrupt routine addresses	IN	01_{16}	; INPUT DATA FROM PORT B
.				
.				
001E		EI		; RE-ENABLE INTERRUPT FLAG
001F		RET		; RETURN TO MAIN PROGRAM
.				
.				
XXXX	MAIN	XX		; COMMENCE MAIN PROGRAM

entering the interrupt program. This prevents the interrupt routine from itself being interrupted. The programmer must ensure that the interrupt flag is re-enabled before a return is made to the main program (see instruction 001E in table 9.1). If the microcomputer has a priority interrupt system (see section 9.8), the interrupt flag must be re-enabled at an early point in the interrupt routine to allow higher-order interrupts to have precedence.

The interrupt routine in table 9.1 calls for data to be input from port B; the remainder of the interrupt routine (not shown) depends on the operations to be carried out on the data from port B. The routine is completed by an EI instruction, followed by a RETurn to the interrupted program instruction.

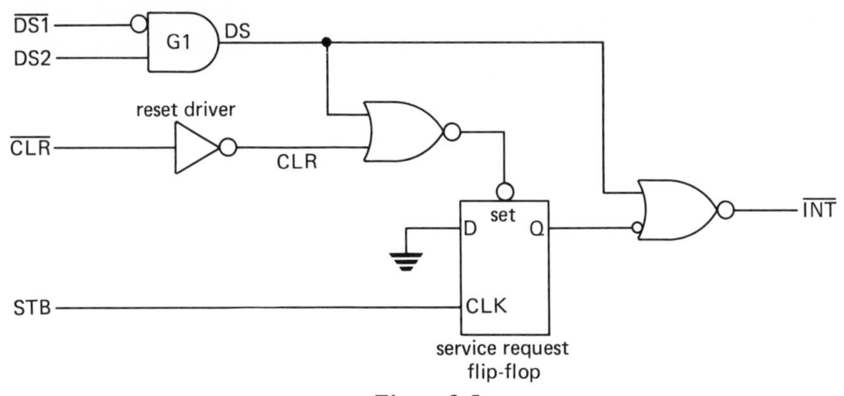

Figure 9.5

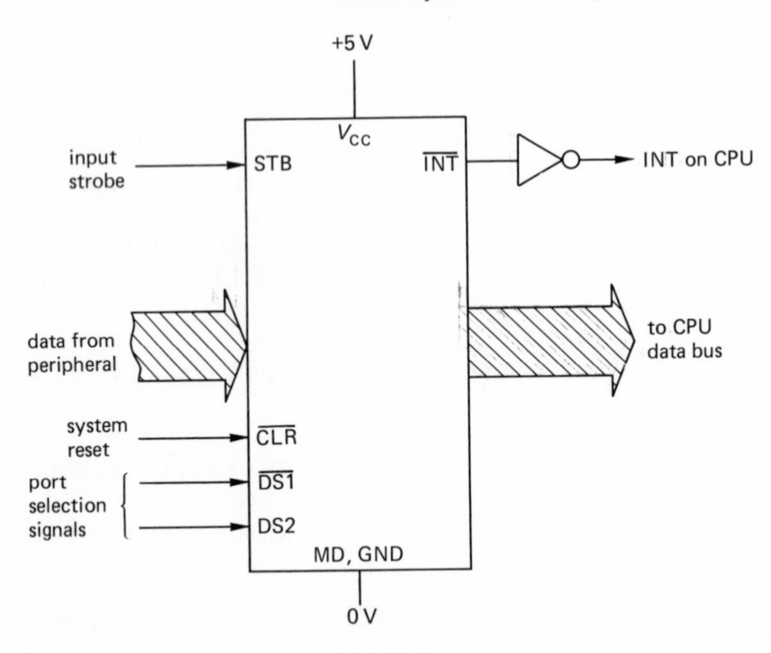

Figure 9.6

Many I/O ports have internal logic circuitry which allows a strobe from the system to generate an interrupt signal via the I/O port. An example of this kind of port is the 8212 chip. A simplified version of an 8212 I/O port without an interrupt signal generating ability was described in chapter 3. The circuit in figure 9.5 shows the additional circuitry needed in order to generate an interrupt signal; the combined circuitry of figures 3.25 and 9.5 is that of a practical I/O port.

Gate G1 in figure 9.5, and the reset driver inverting gate are also part of the basic I/O port in figure 3.25. The 8212 I/O port produces an active low signal at the $\overline{INT}$ output connection either if signal DS is high OR if output Q from the service request flip-flop is low. Since the CPU described in this book needs an active-high signal to activate the interrupt line, an inverting buffer must be used between the $\overline{INT}$ output from the 8212 port and the INT input of the CPU.

When a logic '0' is applied to the $\overline{CLR}$ input of the 8212 port, it sets output Q of the service request flip-flop to logic '1'; the port is then in its non-interrupting state. When a logic '1' is applied to the STB (STROBE) input, it resets output Q of the service request flip-flop to logic '0', forcing the $\overline{INT}$ output line low. A block diagram of a port operating in this manner is shown in figure 9.6. The port shown is described as an **interrupting input port.**

9.7 Interrupt Program Organisation

A flowchart illustrating the general procedure for processing an interrupt is shown

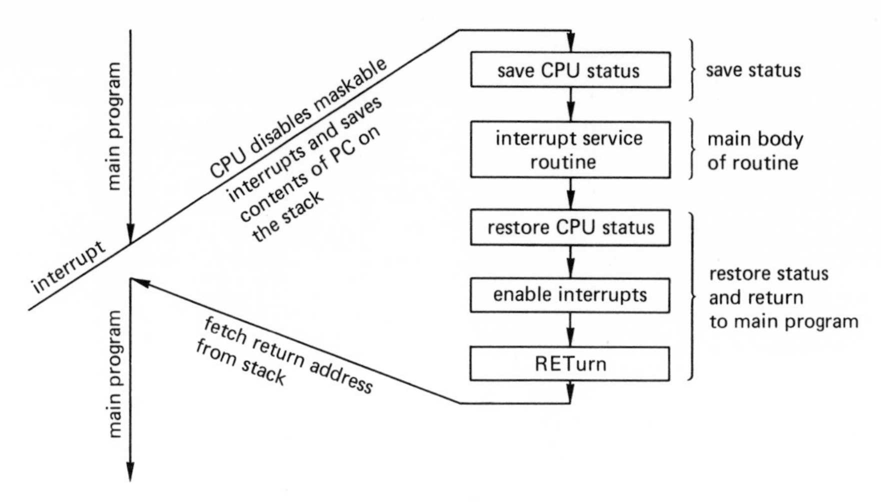

Figure 9.7

in figure 9.7. It follows the general pattern outlined for a subroutine, namely

(1) The contents of the registers (A, FR, B, C, D, E, H, L) should, if necessary, be saved on the stack (this is known as **saving the CPU status**).
(2) The interrupt service routine is executed.
(3) The contents of the registers are restored, the interrupt flag is re-enabled, and an orderly return is made to the main program.

The reader should note that step 1 above is optional, since it may not be necessary in a particular routine to save the contents of the registers (see, for example, table 9.1). A feature of microprocessors is that maskable interrupts are disabled when a transfer is made from the main program to the interrupt routine and, at the same time, the contents of the program counter are saved on the stack.

The memory map adopted by the programmer depends on many factors, and one possible map for an interrupt-driven system is shown in figure 9.8.

9.8 Priority Interrupts

In a **priority interrupt system**, each interrupting device has a specified priority, and a high-priority interrupt takes precedence over one with lower priority, even if the lower-priority interrupt is currently being serviced. A chart illustrating a two-priority interrupt-driven system in which interrupting device B has priority over interrupting device A is shown in figure 9.9.

Priority interrupt circuits vary from a simple system constructed from discrete logic gates up to complex circuits containing MSI chips (an example of the latter is the 8214 Priority Interrupt Control Unit).

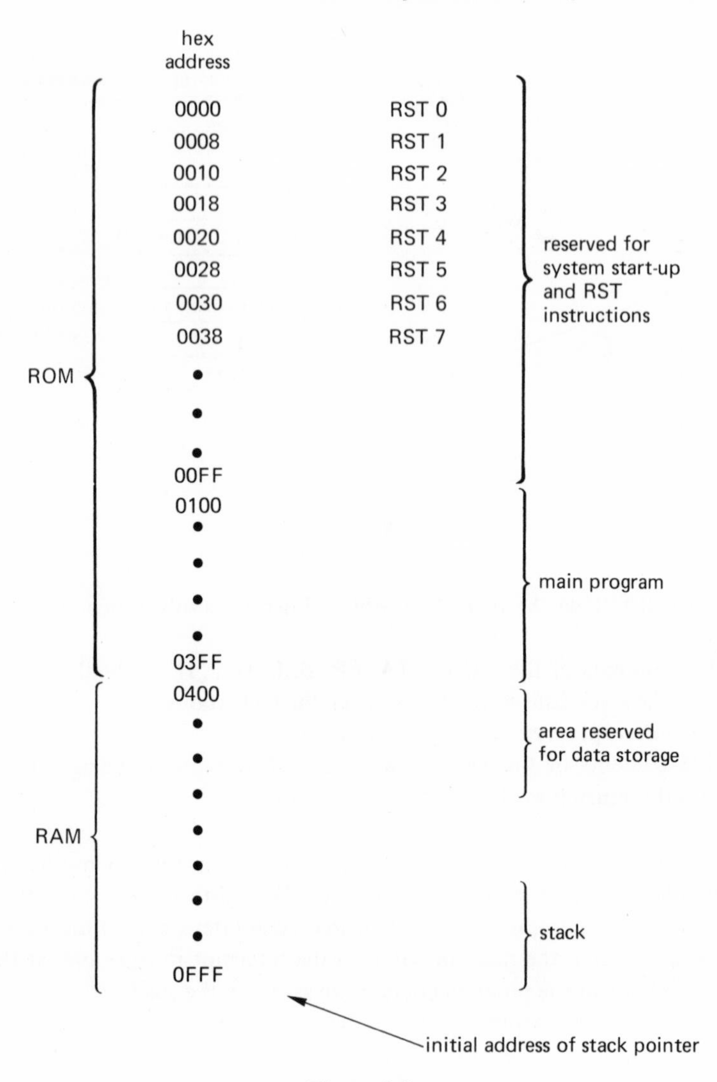

Figure 9.8

One circuit which uses an 8-line-to-3-line priority encoder (such as a 74148 TTL chip) together with an 8212 interrupt instruction port is shown in figure 9.10. The reader is referred to section 3.19 where the 74148 priority encoder is described; note that the inputs are active low. With the connections shown, a logic '0' on input S produces the RST 7 restart code at the output of the encoder. Similarly, a low signal on input Z produces the RST 0 code. Moreover, the priority encoder is arranged so that the signal applied to input I_0 (signal S) has the lowest priority, while that applied to input I_7 (signal Z) has the highest priority. Thus RST 0 has the highest priority and RST 7 has the lowest priority.

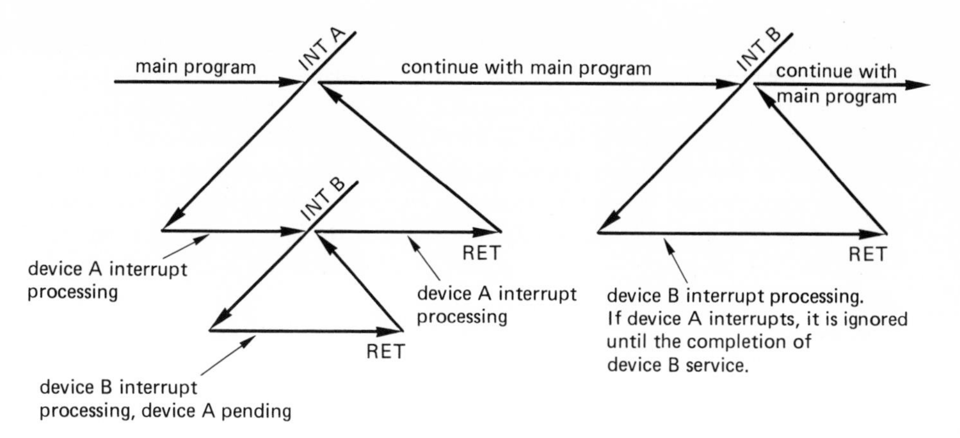

Figure 9.9

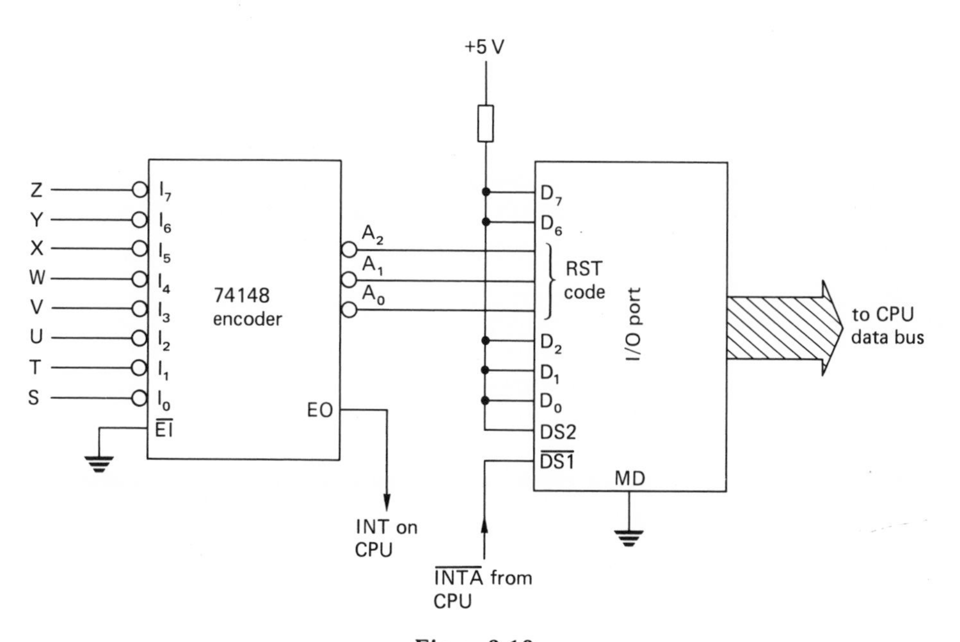

Figure 9.10

9.9 Timers

Many applications involving microprocessors involve sampling data on a regular basis. For example, a CPU installed in a car may not only have to function as a clock, but also calculate the rate of fuel consumption as well as the distance covered during the journey and many other quantities. One method of using the CPU as a clock in this application would be to cause it to execute a timing subroutine, which increments the clock reading every minute. However, such a procedure demands the exclusive use of the CPU, and it would not be available for other uses.

An alternative method of measuring time (and one which is very popular) is to use an external timing device (such as the 8253 chip) which is driven by the CPU clock, and is therefore very accurate. This device functions as a multi-purpose timer/counter element which is treated as an I/O port by the system software. Such a timer can drive the CPU via an interrupt signal at the appropriate moment. As a result, the CPU does not need to keep track of the time as this is done by a support chip.

In general there are two principal applications of timers, namely

(1) As an interval timer
(2) As a 'real-time' clock

An **interval timer** is used to provide output pulses at precise time intervals between operations. An example of an interval timer is a washing machine controller. When the washing machine is full of water at the correct temperature, the timer allows the machine to perform various sequences such as washing, rinsing, draining, for precise intervals of time.

A **real-time clock** provides output pulses at precise intervals of time. The automobile clock mentioned above is an example of this kind. Another example of a real-time clock is in a central heating system, where the 'clock' provides signals at specific times during the day in order to turn the heating system on and off.

PROBLEMS

9.1 Discuss the advantages and drawbacks associated with an interrupt-driven system when compared with a non-interrupt-driven system.

9.2 List a number of applications using (a) non-maskable interrupts, (b) maskable interrupts.

10 A Typical Instruction Set

10.1 Types of Instruction

The instruction set of a microprocessor is the repertoire of commands or instructions which the programmer uses to cause the CPU to produce known responses.

Broadly speaking, the instruction set of any CPU can be divided into four categories, which are respectively concerned with data manipulation, data transfer, program manipulation and status management. Each type is described in detail in this chapter. Other subdivisions of the above categories are possible, and the reader will find many examples of them in manufacturers literature.

From earlier work, the reader will appreciate that the CPU can only accept data in a coded form, that is in the form of a series of 1's and 0's, in other words in **machine code**. To enable the programmer to write a program quickly and in a way that may be read and understood by others, an assembly language is used. In an **assembly language**, a mnemonic is assigned to each instruction that the CPU can execute. The majority of assembly languages are machine dependent, that is each one is designed for use with only one type of CPU.

A program written in an assembly language is known as a **source program** (see, for example, table 8.7), which is later converted into an **object program** that is in machine code instructions and is ready for execution. Each assembly language instruction corresponds to one or more bytes of machine code, the conversion being completed either by hand or by an **assembler program**.

10.2 Instruction and Data Format

Depending on the type of CPU, each computer word is organised either as 4, or 8 or 16 contiguous bits, with an 8-bit word being most popular. Each program instruction consists of one, two, three or four bytes, multiple bytes being stored in successive memory locations. Typical instruction formats are given in table 10.1.

The **op-code** or **operation code** is the part of the instruction which tells the CPU what operation must be performed next. In order briefly to introduce the reader to the four types of instruction in table 10.1, an example of each type is given in table 10.2 (see also table 8.6 for details of some of them).

The range of instructions dealt with by the CPU described in this book are shown in table 10.3 (details are given later). Since an 8-bit word can represent any value up to FF_{16} (256_{10}), the first byte of the op-code can deal with 256_{10} instruc-

Table 10.1

	Type of instruction			
	1-byte	2-byte	3-byte	4-byte
First byte	op-code	op-code	op-code	op-code
Second byte		data or address	data or address	op-code
Third byte			data or address	displacement or data or address
Fourth byte				data or address

Table 10.2

	Op-code	Mnemonic	Comment
1-byte	C9	RET	Return to main program
2-byte	0E 68	MVI C,68_{16}	Load register C with 68_{16}
3-byte	C2 14 01	JNZ,0114_{16}	Jump if not zero to address 0114
4-byte	DD 36 F1 24	LD(IX–15),24_{16}	Load 24_{16} into a location using indexed addressing

tions, not all of which are used in the instruction set adopted here. In the case of a 2-byte instruction, the second byte may either be data or it may be an address (as, for example, in the case of addressing an I/O port which has accumulator I/O addressing). The second and third bytes of a 3-byte instruction could, for example, contain the address of a memory-mapped device (where sixteen bits are needed to specify the address of the location). In the case of a 4-byte instruction, the op-code is defined by the first two bytes; that is, the first byte of the op-code has 256_{10} possible variations (not all of which are used) which are provided by the second byte of the instruction (see table 10.4).

The instructions given in table 10.3 are basically those of the 8080 CPU, but have been extended to deal with indexed addressing (see also section 10.9).

10.3 Microinstructions and Microprograms

Thus far in the book, it has been assumed that a single instruction is executed in

Table 10.3 Instructions associated with the first op-code byte

Least significant hex character (columns); Most significant hex character (rows)

	0	1	2	3	4	5	6	7	8	9	A	B	C	D	E	F
0	NOP 1	LXI B 3	STAX B 1	INX B 1	INR B 1	DCR B 1	MVI B 2	RLC 1		DAD B 1	LDAX B 1	DCX B 1	INR C 1	DCR C 1	MVI C 2	RRC 1
1		LXI D 3	STAX D 1	INX D 1	INR D 1	DCR D 1	MVI D 2	RAL 1	JR 2	DAD D 1	LDAX D 1	DCX D 1	INR E 1	DCR E 1	MVI E 2	RAR 1
2	JRNZ 2	LXI H 3	SHLD 3	INX H 1	INR H 1	DCR H 1	MVI H 2	DAA 1	JRZ 2	DAD H 1	LHLD 3	DCX H 1	INR L 1	DCR L 1	MVI L 2	CMA 1
3	JRNC 2	LXI SP 3	STA 3	INX SP 1	INR M 1	DCR M 1	MVI M 2	STC 1	JRC 2	DAD SP 1	LDA 3	DCX SP 1	INR A 1	DCR A 1	MVI A 2	CMC 1
4	MOV B,B 1	MOV B,C 1	MOV B,D 1	MOV B,E 1	MOV B,H 1	MOV B,L 1	MOV B,M 1	MOV B,A 1	MOV C,B 1	MOV C,C 1	MOV C,D 1	MOV C,E 1	MOV C,H 1	MOV C,L 1	MOV C,M 1	MOV C,A 1
5	MOV D,B 1	MOV D,C 1	MOV D,D 1	MOV D,E 1	MOV D,H 1	MOV D,L 1	MOV D,M 1	MOV D,A 1	MOV E,B 1	MOV E,C 1	MOV E,D 1	MOV E,E 1	MOV E,H 1	MOV E,L 1	MOV E,M 1	MOV E,A 1
6	MOV H,B 1	MOV H,C 1	MOV H,D 1	MOV H,E 1	MOV H,H 1	MOV H,L 1	MOV H,M 1	MOV H,A 1	MOV L,B 1	MOV L,C 1	MOV L,D 1	MOV L,E 1	MOV L,H 1	MOV L,L 1	MOV L,M 1	MOV L,A 1
7	MOV M,B 1	MOV M,C 1	MOV M,D 1	MOV M,E 1	MOV M,H 1	MOV M,L 1	HLT 1	MOV M,A 1	MOV A,B 1	MOV A,C 1	MOV A,D 1	MOV A,E 1	MOV A,H 1	MOV A,L 1	MOV A,M 1	MOV A,A 1
8	ADD B 1	ADD C 1	ADD D 1	ADD E 1	ADD H 1	ADD L 1	ADD M 1	ADD A 1	ADC B 1	ADC C 1	ADC D 1	ADC E 1	ADC H 1	ADC L 1	ADC M 1	ADC A 1
9	SUB B 1	SUB C 1	SUB D 1	SUB E 1	SUB H 1	SUB L 1	SUB M 1	SUB A 1	SBB B 1	SBB C 1	SBB D 1	SBB E 1	SBB H 1	SBB L 1	SBB M 1	SBB A 1
A	ANA B 1	ANA C 1	ANA D 1	ANA E 1	ANA H 1	ANA L 1	ANA M 1	ANA A 1	XRA B 1	XRA C 1	XRA D 1	XRA E 1	XRA H 1	XRA L 1	XRA M 1	XRA A 1
B	ORA B 1	ORA C 1	ORA D 1	ORA E 1	ORA H 1	ORA L 1	ORA M 1	ORA A 1	CMP B 1	CMP C 1	CMP D 1	CMP E 1	CMP H 1	CMP L 1	CMP M 1	CMP A 1
C	RNZ 1	POP B 1	JNZ 3	JMP 3	CNZ 3	PUSH B 1	ADI 2	RST 0 1	RZ 1	RET 1	JZ 3	JMPI 3	CZ 3	CALL 3	ACI 2	RST 1 1
D	RNC 1	POP D 1	JNC 3	OUT 2	CNC 3	PUSH D 1	SUI 2	RST 2 1	RC 1		JC 3	IN 2	CC 3	IXR 3/4	SBI 2	RST 3 1
E	RPO 1	POP H 1	JPO 3	XTHL 1	CPO 3	PUSH H 1	ANI 2	RST 4 1	RPE 1	PCHL 1	JPE 3	XCHG 1	CPE 3		XRI 2	RST 5 1
F	RP 1	POP PSW 1	JP 3	DI 1	CP 3	PUSH PSW 1	ORI 2	RST 6 1	RM 1	SPHL 1	JM 3	EI 1	CM 3		CPI 2	RST 7 1

Note: the number in the lower right-hand corner of each cell is the number of bytes in the instruction.

Table 10.4

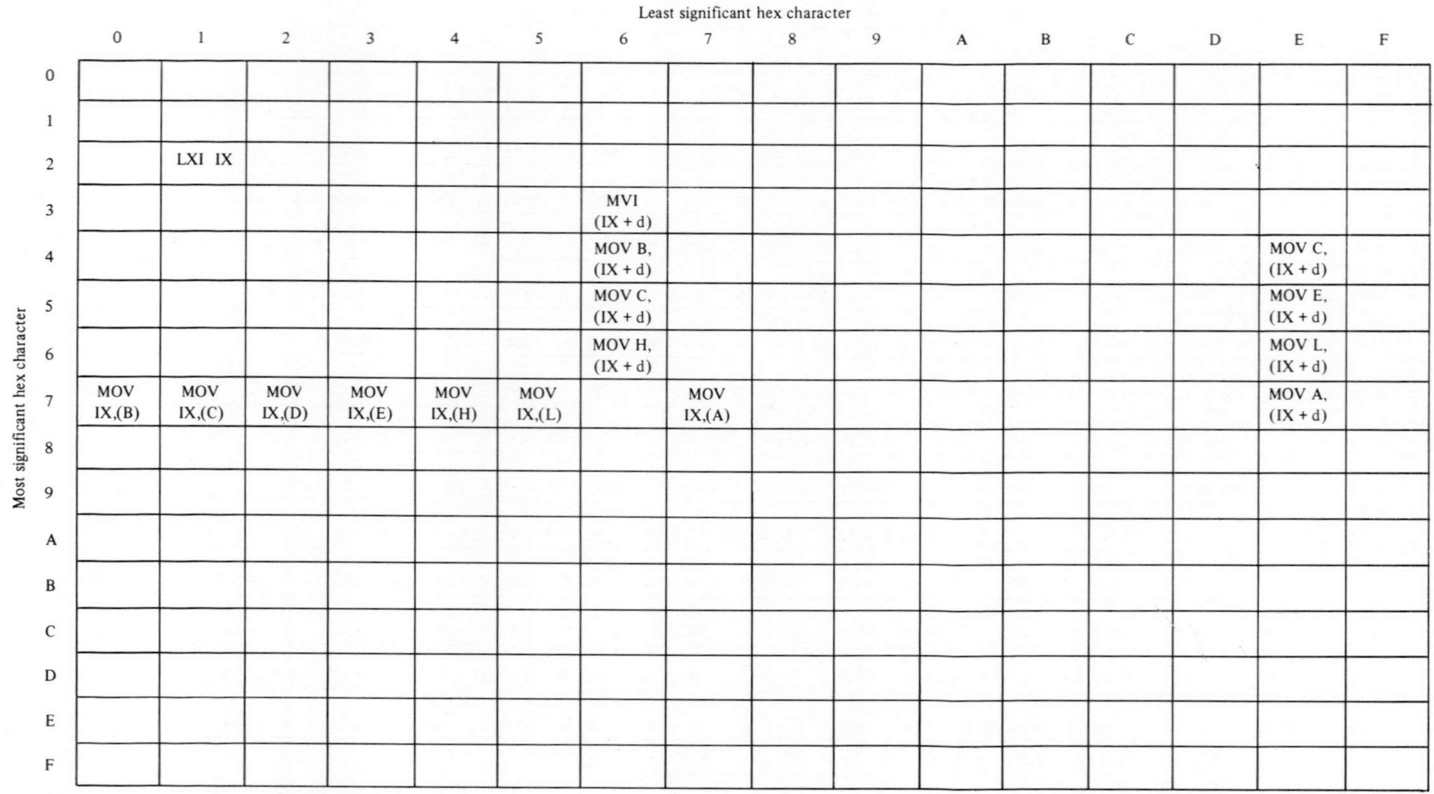

Most significant hex character \ Least significant hex character	0	1	2	3	4	5	6	7	8	9	A	B	C	D	E	F
0																
1																
2		LXI IX														
3							MVI (IX + d)									
4							MOV B, (IX + d)								MOV C, (IX + d)	
5							MOV C, (IX + d)								MOV E, (IX + d)	
6							MOV H, (IX + d)								MOV L, (IX + d)	
7	MOV IX,(B)	MOV IX,(C)	MOV IX,(D)	MOV IX,(E)	MOV IX,(H)	MOV IX,(L)		MOV IX,(A)							MOV A, (IX + d)	
8																
9																
A																
B																
C																
D																
E																
F																

A, B, C, D, E, H and L are registers: d is a displacement (hex)

one step. In fact this is not the case, and each instruction really comprises a series of **microinstructions**; for example, the execution of an LDI (Load Immediate) instruction consists of executing six microinstructions. The great majority of CPUs are microprogrammed at the factory; however, if the control unit can be modified by the user, then the CPU is said to be **microprogrammable**. A microprogrammable microprocessor allows the user either to add or to replace microprograms in the basic model.

10.4 Macro-assemblers and Pseudo-operations

A **macro** is a sequence of one or more instructions which can be given a name (for example, the time delay program described in chapter 9 could represent a macro called TIME). Whenever the name is written in the source program, the assembler automatically expands the macro so that the programmer need not write it out in full. The majority of macros consist of only a few instructions, and perhaps the most powerful use of macros is in I/O routines.

A **pseudo-operation** or **pseudo-op** is an assembly language directive which does not generate an object code (that is, 0's and 1's) during the assembling process. Each assembler has its own range of pseudo-ops, and the reader is advised to study the function of each of them in a microprocessor with which he is familiar. Two pseudo instructions which are widely used are ORG (ORiGin) and END. The first line of the source program could begin

ORG 0F20H

which causes the CPU to store the next instruction (which is the first instruction in the program proper) in the hexadecimal address (signified by H at the end of the pseudo-op) 0F20. The final line in the program contains the word END, which informs the CPU via the assembler that it has dealt with the entire program.

10.5 Addressing Modes

The number and variety of addressing modes which the CPU can use provide an indication of the effectiveness of the CPU. Addressing methods are many and varied, and popular types include

(1) Direct addressing
(2) Indirect addressing
(3) Immediate addressing
(4) Indexed addressing
(5) Relative addressing
(6) Register direct addressing
(7) Register indirect addressing
(8) Stack addressing or auto-indexed addressing

10.6 Direct Addressing

In this case, the address of the operand is part of the instruction. Depending on the type of instruction to be executed, this method of addressing requires either a 2-byte (**page zero**) or a 3-byte (**full direct addressing**) instruction, as follows

byte 1 op-code
byte 2 address within page
byte 3 page address

For example, the instruction IN 01 is a 2-byte (page zero) direct addressing instruction which causes data to be loaded into the accumulator from input port 01 (which is at location 01 on page zero, that is, at address 0001_{16}).

The 3-byte instruction STA 0242_{16} (full direct addressing) STores the Accumulator contents in location 42_{16} on page 02_{16}. Full direct addressing is sometimes described as **extended addressing**, since all sixteen bits are used to define the address of the location.

10.7 Indirect Addressing

In this method of addressing, the address part of the instruction refers to another location which contains a second address; the second address specifies the operand.

As an example of this addressing mode, consider the 3-byte JMPI instruction (JuMP by Indirect addressing) used by the CPU in this book, having the op-code CB (see table 10.3). The steps carried out are shown in figure 10.1. The JMPI op-code in location 0021 causes the control unit to interpret the byte in location 0022 as the low byte of the address at which the data is stored, and it interprets the contents of location 0023 as the high byte of the address at which the data is stored. Program control is therefore transferred to address 0F60. The JMPI instruction then results in the CPU transferring the contents of location 0F60 to the low byte of the program counter, and the contents of location 0F61 to the high byte of the PC. The next instruction to be executed is therefore at location BCOA.

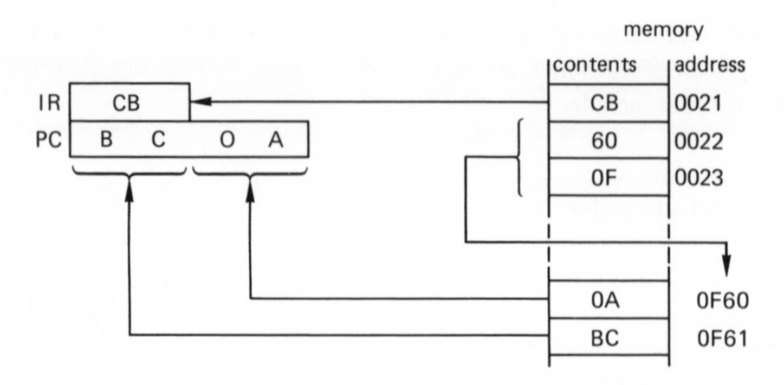

Figure 10.1

10.8 Immediate Addressing

In this method of addressing, the operand is part of the instruction (usually the second byte). An example is the MVI r,data (MoVe Immediate a byte of data into register r); the MVI B,$5A_{16}$ instruction causes the hex value 5A to be moved immediately into register B.

Certain instructions allow a 16-bit binary value to be moved into a register pair. An example is the LXI rp instruction; the instruction LXI SP,$FF20_{16}$ causes the stack pointer address to be initialised at $FF20_{16}$.

10.9 Indexed Addressing

In this case the address in the instruction is modified by the contents of an **index register**; the actual address or **effective address** being given by the combination of the address part of the instruction and the contents of the index register. Indexed addressing is very useful where a program deals with lists and tables of data stored in memory. Indexed addressing is implemented here either in a 3-byte or a 4-byte format, as shown in table 10.5.

Table 10.5

	3-byte instruction	4-byte instruction
First byte	Op-code	Op-code
Second byte		
Third byte	Displacement	Displacement or data
Fourth byte		Data

In order to provide negative displacements, the 'displacement' in table 10.5 is given in radix complement notation (see chapter 2 and also the example which follows). When using indexed addressing, it is first necessary to load the index register IX (see figure 8.2) with a 16-bit address using an instruction such as DD 21 00 0F as follows

first byte	DD	op-code
second byte	21	
third byte	00	lower address byte
fourth byte	0F	upper address byte

The net result is that the index register IX stores a base address of $0F00_{16}$. The above sequence of instructions should ideally be inserted at an early point in the program. Data can be loaded into a destination using indexed addressing by means of a 4-byte instruction as, for example, by

MVI (IX + d), n

This instruction causes (n) to be MoVed Immediately into an address given by the sum of the contents of IX and the displacement (d). If the base address of the index register is 0F00, then the number $1A_{16}$ is loaded into the effective address $0F56_{16}$ by means of the instruction DD 36 56 1A as follows

DD 36	op-code
56	displacement = +56
1A	data to be stored

The effective address is calculated as follows

0F 00	base address
(00)56	displacement (+56)
0F56	effective address

In the above calculation, the two most significant zeros of the displacement are shown in brackets since they do not appear in the instruction itself, but are implied in its execution. The displacement is positive if its value is in the range 00_{16} to $7F_{16}$ (0_{10} to 127_{10}), and is negative if it is in the range 80_{16} to FF_{16} (-128_{10} to -1_{10}. For example, if we wish to store $1A_{16}$ in address 0ED1 using a base address 0F00, the instruction needed is

DD 36	op-code
D1	displacement (-47_{10} or $-2F_{16}$ in complement form)
1A	data

The effective address is calculated as follows

0F 00	base address
(FF)D1	displacement in 2's complement form
1 ← 0E D1	effective address
↑ overflow (lost)	

Once again, the two most significant numbers in the displacement (F's since it is a negative value) are enclosed in brackets since they are implicit in the execution of the instruction.

10.10 Relative Addressing

In this method of addressing, the address specified in the instruction is a displacement from a base address. The base address may be the contents of the program counter itself, or it may be stored in a base register.

Relative addressing allows a complete program to be moved from one area of memory and relocated in another area of memory without change to the program. Such a program is said to be **relocatable.**

The type of relative addressing used in this book is limited to JUMPs relative to the value stored in the program counter. The general form of the instruction is

first byte - op-code
second byte - displacement (stored in complement notation)

The JR instructions (Jump Relative) in table 10.6 are also shown in table 10.3.

Table 10.6

Machine code	Mnemonic	Condition for jump
18	JR	Unconditional jump
20	JRNZ	Jump if the zero flag is not set
28	JRZ	Jump if the zero flag is set
30	JRNC	Jump if the carry flag is not set
38	JRC	Jump if the carry flag is set

To understand the calculation of the jump 'displacement', the reader must first appreciate the mechanics of the operation of the program counter. When fetching each byte of information (either op-code or displacement), the CPU increments the contents of the program counter; this ensures that the PC points to the next byte to be fetched.

Table 10.7

Symbolic address		Contents of PC	Machine code	Op-code	Comment
A		0200	28	JRZ 04	Jump if zero flag set
A + 1		0201	04		
A + 2	┌	0202 ←			Value in PC after JRZ instruction
A + 3	jump	0203			
A + 4	jump	0204			
A + 5	jump	0205			
A + 6	└→	0206			Effective address

Consider the case illustrated in table 10.7. Here a 2-byte JRZ instruction appears at the head of the table, the displacement of +4 being given in the second byte of the instruction; after the CPU has fetched both bytes of the instruction, the program counter points at address 0202 (that is, to address A+2). So far as the program 'jump' is concerned, the CPU regards address 0202 (that is, A + 2) as address 'zero' and, if the zero status flag is set, program control is transferred to address 0206 as follows

nominal 'zero' address	02 02	or	A + 2
displacement	(00)04		4
destination (effective address)	02 06		A + 6

Since the displacement is stored in complement form, this method of addressing deals with positive jumps up to $7F_{16}$ (127_{10}) and negative jumps up to 80_{16} (-128_{10}). Allowing for the two bytes of the jump instruction, the effective displacement from the JR op-code address (0200_{16} in this case) is in the range $+129_{10}$ to -126_{10}. This is illustrated in figure 10.2, in which a program jump can be made relative to the first byte of the JR instruction to any address in the shaded area. The destination address can be calculated from the following formula.

destination address = JR address + displacement + 2

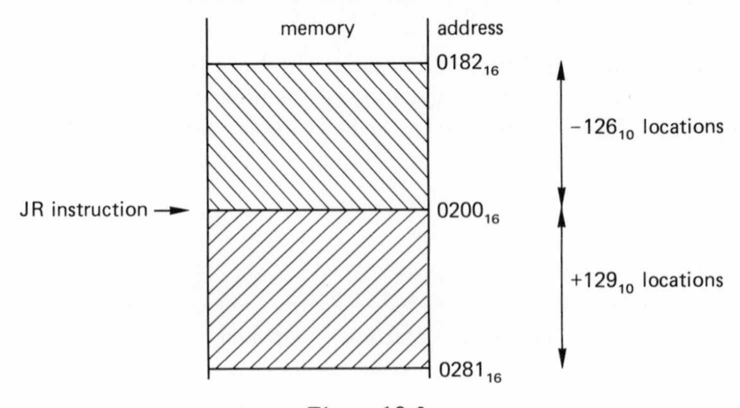

Figure 10.2

If the displacement in the above example had been $F8_{16}$ (equivalent to -8_{10}), then the destination address would be

JR address	02 00
displacement	(FF)F8
partial sum	01 F8
+2	2
destination address	01 FA

CPUs which incorporate relative addressing include many more instructions than are listed in table 10.6. For example they include MOVE (or LOAD) instructions, etc.

When using relative addressing, the reader is advised to study the instruction manual of the CPU he is using very closely to find if there are any apparent variations in the use of the addressing method. For example, when a PC relative displacement of 80_{16} is used with the National 8060 CPU, the actual displacement is not necessarily 80_{16} (-128_{10}) but is the value given by the contents of a special user-accessible register.

10.11 Register Direct Addressing

This is generally the same as memory direct addressing (see section 10.6), but the address is a register rather than a memory location. An example of an instruction of this kind is one which moves data from one register to another. For example, the 1-byte instruction

MOV C,A (op-code 4F)

transfers the contents of register A (the accumulator) into register C.

10.12 Register Indirect Addressing

In this method of addressing, the address part of the instruction refers to a register pair which contains a second address; the second address specifies the operand (this is much the same as memory indirect addressing - see section 10.7).

For example, the 1-byte instruction LDAX B (op-code 0A) causes the accumulator to be loaded with data from the location whose address is stored in register pair B,C. The execution of this instruction is illustrated in figure 10.3; here the register pair B,C contains address 0256, hence the contents of this address (12_{16} in

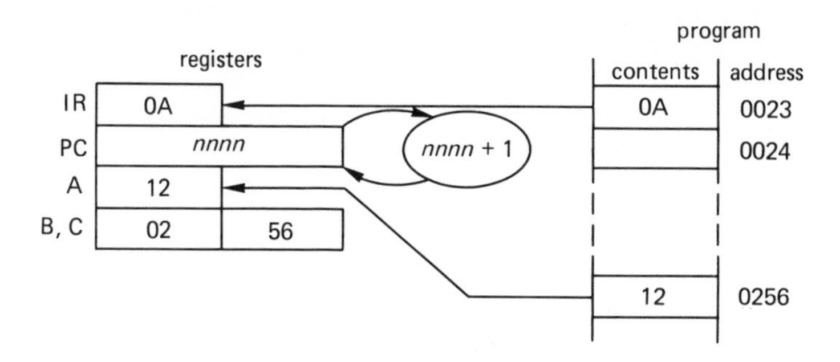

Figure 10.3

the example shown) are loaded into the accumulator. The content of the program counter is incremented by unity on the completion of this instruction.

10.13 Stack Addressing

Stack addressing is implemented via the address stored in the stack pointer. The reader is referred to chapter 8 for further details on this method. This mode is also known as auto-indexed addressing.

10.14 Instruction Categories

The instructions used in this book have been divided into four broad categories, namely

(1) Data manipulation instructions
(2) Data transfer instructions
(3) Program manipulation instructions
(4) Status management instructions

10.15 Symbols and Abbreviations Used when Describing Instructions

The following are used in association with the instructions described in this chapter.

r, r1, r2 one of the registers A, B, C, D, E, H, L, IX

rp a register pair as follows
B represents the pair B,C
D represents the pair D,E
H represents the pair H,L
PSW represents the pair A,FR

rh the high-order register of a designated register pair as follows
B for rp B,C
D for rp D,E
H for rp H,L

rl the low-order register of a designated register pair as follows
C for rp B,C
E for rp D,E
L for rp H,L

PC 16-bit program counter
SP 16-bit stack pointer
IX 16-bit index pointer register
Z zero flag
S sign flag
P parity flag
C carry flag
A_C auxiliary carry flag

( )	'the contents of' a register or memory
←	'is transferred to'
↔	'is interchanged with'
∧	logical AND
V	logical OR
⊻	EXCLUSIVE OR
+	addition
-	2's complement subtraction
*	multiplication

10.16 Data Manipulation Instructions

These instructions usually (but not always) cause data to be modified in some way. They can be broadly categorised into

(1) Arithmetic instructions
(2) Comparison instructions
(3) Logical instructions
(4) Shift and rotate instructions

The above categories of instructions are listed in table 10.8

Arithmetic instructions

This group of instructions performs arithmetic operations on data in registers and memory, and may affect the state of the status flags. The basic operations are ADD, ADD WITH CARRY, DECIMAL ADJUST ACCUMULATOR, SUBTRACT, SUBTRACT WITH BORROW, INCREMENT, DECREMENT. Each type is described in the following.

The basic 1-byte ADD instruction using register direct addressing is written in mnemonic form as ADD r (ADD the contents of a register to the accumulator). That is

$$\text{ADD r: (A)} \leftarrow \text{(A)} + \text{(r)}$$

The operation above is read from right to left as follows

Add the contents of register r to the contents of the accumulator,
the result being stored in the accumulator.

This type of notation is widely used when describing CPU instructions. Examples of the above instruction appear in table 10.8 and include ADD B (op-code 80), ADD C (op-code 81), ADD D (op-code 82) etc. If, for example, the register specified in the instruction stores $3A_{16}$ and the accumulator initially contained 92_{16}, then after execution of the instruction the register still stores $3A_{16}$ and the accumulator stores $(3A + 92)_{16} = CC_{16}$.

The 1-byte ADD M instruction which uses register indirect addressing, adds the contents of a memory location (M) which is specified by the address stored in the

Table 10.8 Data manipulation instructions

Least significant hex character (columns); Most significant hex character (rows)

	0	1	2	3	4	5	6	7	8	9	A	B	C	D	E	F
0				INX B	INR B	DCR B		RLC		DAD B		DCX B	INR C	DCR C		RRC
1				INX D	INR D	DCR D		RAL		DAD D		DCX D	INR E	DCR E		RAR
2				INX H	INR H	DCR H		DAA		DAD H		DCX H	INR L	DCR L		CMA
3				INX SP	INR M	DCR M				DAD SP		DCX SP	INR A	DCR A		
4																
5																
6																
7																
8	ADD B	ADD C	ADD D	ADD E	ADD H	ADD L	ADD M	ADD A	ADC B	ADC C	ADC D	ADC E	ADC H	ADC L	ADC M	ADC A
9	SUB B	SUB C	SUB D	SUB E	SUB H	SUB L	SUB M	SUB A	SBB B	SBB C	SBB D	SBB E	SBB H	SBB L	SBB M	SBB A
A	ANA B	ANA C	ANA D	ANA E	ANA H	ANA L	ANA M	ANA A	XRA B	XRA C	XRA D	XRA E	XRA H	XRA L	XRA M	XRA A
B	ORA B	ORA C	ORA D	ORA E	ORA H	ORA L	ORA M	ORA A	CMP B	CMP C	CMP D	CMP E	CMP H	CMP L	CMP M	CMP A
C							ADI								ACI	
D							SUI								SBI	
E							ANI								XRI	
F							ORI								CPI	

register pair HL, to the contents of the accumulator; the result is stored in the accumulator. For example, if the accumulator stores 92_{16}, and the HL register pair store an address 0B01, then the ADD M instruction (op-code 86) causes the contents of address 0B01 to be added to 92_{16}, the result being stored in the accumulator. The ADD M instruction is represented as follows

$$\text{ADD M: (A)} \leftarrow \text{(A)} + \text{((HL))}$$

where ((HL)) represents the contents of the memory location addressed by HL.

The 1-byte ADd with Carry (ADC) instructions (that is, ADC r and ADC M) are generally similar to the ADD r and ADD M instructions described above, with the exception that the condition of the carry status flag is added to the least significant bit of the sum. The instructions are represented by

$$\text{ADC r (ADd register with Carry): (A)} \leftarrow \text{(A)} + \text{(r)} + \text{(C)}$$
$$\text{ADC M (ADd Memory with Carry): (A)} \leftarrow \text{(A)} + \text{((HL))} + \text{(C)}$$

The 2-byte ADI (ADd Immediate - op-code C6) results in the data in the second byte of the instruction being added to the contents of the accumulator; the result is stored in the accumulator. The ADI instruction results in the following operation

$$\text{ADI data: (A)} \leftarrow \text{(A)} + \text{(byte 2 of instruction)}$$

Similarly, the ACI instruction (Add with Carry Immediate - op-code CE) results in

$$\text{ACI data: (A)} \leftarrow \text{(A)} + \text{(byte 2 of instruction)} + \text{(C)}$$

The DAA instruction (Decimal Adjust Accumulator - op-code 27) is used in cases where the 8-bit number in the accumulator is adjusted to form two 4-bit BCD characters. The reason for this instruction was outlined in chapter 2. This instruction is needed, for example, following the addition of two decimal numbers as follows

```
MVI A   77   ;(A) ← 77
ADI     08   ;(A) ← (A) + 08
DAA          ; convert the result to decimal
```

The operations in the above program are explained below. The MVI A instruction was introduced earlier in the book and is also described later in this section. It results in the hex number 77 being MoVed Immediately into the accumulator; the reader will observe that since hex numbers in the range zero to 9 are equivalent to decimal numbers, then we can regard the hex value 77 as a decimal value 77. The ADI 08 instruction causes the CPU to carry out the hex addition (77 + 8) to leave $7F_{16}$ in the accumulator. Finally, the DAA instruction adjusts the value in the accumulator to the decimal sum 85_{10}.

In some cases the programmer has to deal with a 16-bit addition. The instruction set used here allows for a DAD instruction (Double length ADdition); this is also known as **double precision addition.** The result of this instruction is summarised below; it results in the contents of a specified register pair being added to the contents of the register pair HL, the sum being placed in register pair HL.

DAD rp: $(H)(L) \leftarrow (H)(L) + (rh)(r1)$

The various subtract instructions are summarised below; SBB and SBI instructions refer to subtract with borrow.

SUB r: $(A) \leftarrow (A) - (r)$
SUB M: $(A) \leftarrow (A) - ((HL))$
SUI data: $(A) \leftarrow (A) - (\text{byte } 2)$
SBB r: $(A) \leftarrow (A) - (r) - (C)$
SBB M: $(A) \leftarrow (A) - ((HL)) - (C)$
SBI data: $(A) \leftarrow (A) - (\text{byte } 2) - (C)$

In addition to the above arithmetic operations, it is sometimes necessary either to increase the contents of a register or memory location by unity or, alternatively, to decrease the contents by unity. The former is known as incrementing the data and the latter as decrementing the data. These changes are brought about by INR instructions (INcRement) and DCR instructions (DeCRement), respectively. An example of a DCR instruction was given in the time delay sequence in chapter 8. The instructions are summarised below.

INR r: $(r) \leftarrow (r) + 1$
INR M: $((HL)) \leftarrow ((HL)) + 1$
DCR r: $(r) \leftarrow (r) - 1$
DCR M: $((HL)) \leftarrow ((HL)) - 1$

The contents of register pairs can also be incremented or decremented; this allows the programmer to deal with numbers larger than FF_{16} (255_{10}). The register pair operations are

INX rp: $(rh)(rl) \leftarrow (rh)(rl) + 1$
DCX rp: $(rh)(rl) \leftarrow (rh)(rl) - 1$

Comparison instructions

There are three instructions of this kind in our instruction set, and their function is to compare a value specified in the instruction with the contents of the accumulator. When the CPU receives a COMPARE instruction, it subtracts the specified value from the contents of the accumulator and alters the status flags in the light of the result of the subtraction; on completion of the instruction, **the content of the accumulator is left unchanged at its original value**; the result of the subtraction is discarded.

This type of instruction is useful when it is necessary to test if a particular condition has been reached, and it frequently precedes a JUMP, CALL or RETURN instruction. The three types of instruction are respectively CMP r (CoMPare register), CMP M (CoMPare Memory) and CPI data (CoMPare Immediate with data) as follows

CMP r: (A) - (r)
CMP M: (A) - ((HL))
CPI data: (A) - (byte 2)

Logical instructions

The logical instructions which are available to programmers differ slightly between CPUs, but generally include AND, OR, EXCLUSIVE-OR and COMPLEMENT instructions. The data involved is either the contents of a register or a memory, or it may be contained in the program (that is, immediate data). The logic function in the instruction operates on the data and on the accumulator contents, the result of the instruction being left in the accumulator. The AND logical instructions are summarised below

ANA r (AND register): $(A) \leftarrow (A) \wedge (r)$
ANA M (AND Memory): $(A) \leftarrow (A) \wedge ((HL))$
ANI data (AND Immediate): $(A) \leftarrow (A) \wedge (\text{byte } 2)$

The AND instruction is particularly useful for masking one bit or group of bits in the accumulator; this frequently precedes a JUMP, CALL or RETURN instruction. It can also be used to turn bits 'off', that is, to force them to zero. For example, the instruction

ANI FE_{16}

unconditionally sets bit zero (b_0) of the accumulator to zero. Other logical instructions in the instruction set are

ORA r (OR Accumulator with r): $(A) \leftarrow (A) \vee (r)$
ORA M (OR Accumulator with M): $(A) \leftarrow (A) \vee ((HL))$
ORI data (OR Immediate with data): $(A) \leftarrow (A) \vee (\text{byte } 2)$
XRA r (eXclusive oR Accumulator with r): $(A) \leftarrow (A) \veebar (r)$
XRA M (eXclusive oR Accumulator with M): $(A) \leftarrow (A) \veebar ((HL))$
XRI data (eXclusive oR Immediate with data): $(A) \leftarrow (A) \veebar (\text{byte } 2)$

Another logical instruction is the CMA instruction (CoMplement the contents of the Accumulator), as follows

CMA: $(A) \leftarrow (\overline{A})$

This instruction forms the 1's complement of the contents of the accumulator by converting the 1's to 0's and the 0's to 1's; the result is stored in the accumulator.

Shift and rotate instructions

Shift and rotate instructions are frequently used when converting between serial data and parallel data transmission formats (see also chapter 7) and also in multiplication and division routines.

Shift instructions are subdivided into logical shift instructions and arithmetic shift instructions, and are described below. The way in which the shifts are implemented are many and varied, and those described here should be regarded as typical. The reader is advised to consult the assembly language manual of the CPU he is using for details.

Logical shift: the contents of the accumulator are shifted one bit, the empty bit is cleared and the bit shifted out is transferred to the carry flag.

The above sequence of events is illustrated in figure 10.4a. After a *logical shift right,* a '0' is moved into the most significant position of the accumulator and the condition

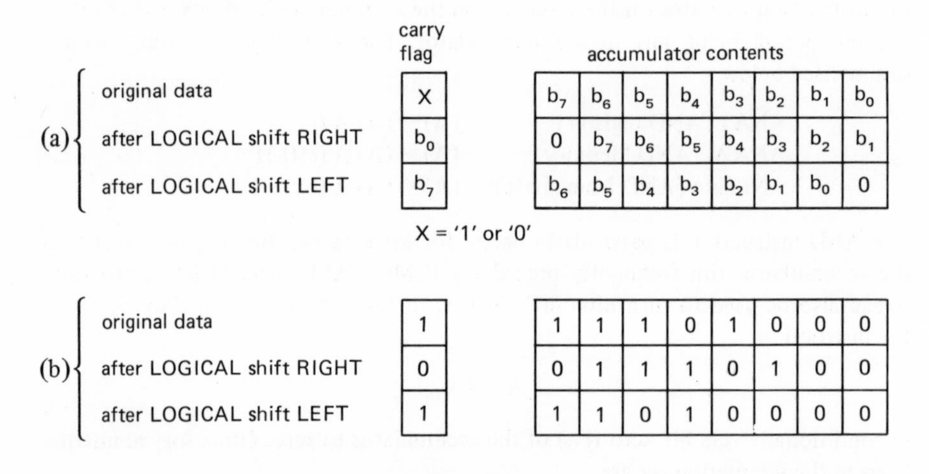

Figure 10.4

of the l.s.b. is copied into the carry flag. After a *logical shift left,* a '0' is moved into the l.s.b. of the accumulator, and the condition of the m.s.b. is copied into the carry flag. Figure 10.4b illustrates these operations for the hex word E8. There are, of course, variations on the above, and in some microprocessors the bit which is shifted out is lost. In general, *a logical shift operation places zeros in the emptied bits.*

arithmetic shift: the contents of the accumulator are shifted one bit, the sign bit (the m.s.b.) being retained. If a bit is emptied, it is cleared, that is, it is set to '0'.

The general principles of arithmetic shift operations are illustrated in figure 10.5a, and examples of arithmetic shift right and left are shown in figure 10.5b. In general, *an arithmetic shift preserves the value of the sign bit (the m.s.b) of the accumulator.*

The instruction set of every CPU has a number of **rotate instructions**, which

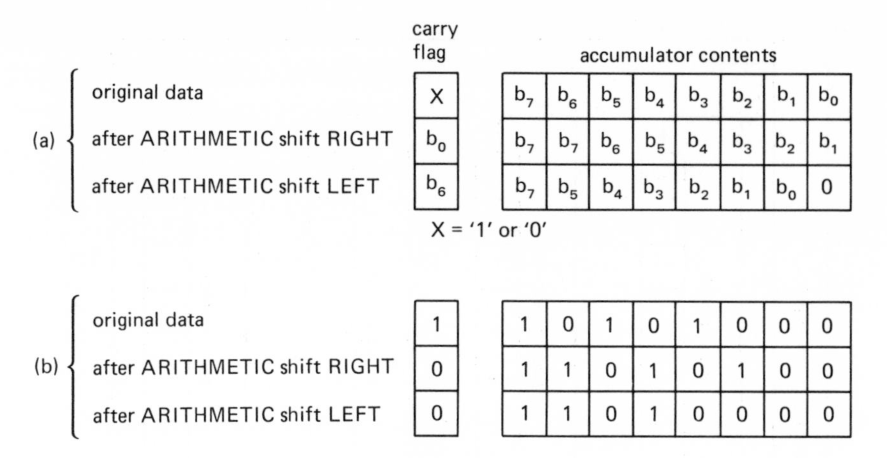

Figure 10.5

result in the data stored in the accumulator being 'rotated' along a circular route in one way or another. The rotate instructions available in the CPU considered here are illustrated in figure 10.6.

The RLC instruction results in the data stored in the accumulator being shifted left, with b_7 being shifted into both b_0 and the C flag.

The RAL and RAR instructions result in the C flag and the accumulator being connected as a 9-bit register, all the bits being shifted circular left or right, respectively.

10.17 Data Transfer Instructions

These instructions result in data being moved from one location in the microcomputer to another. The data is not altered by these instructions. The instructions may be categorised as

(1) Memory transfer operations
(2) Register transfer operations
(3) I/O data transfers
(4) Stack operations

The instructions above are listed in table 10.9.

Memory transfer operations

These instructions control the movement of data between registers and memory locations; the 1-byte instructions in this group are summarised as

MOV r,M: $(r) \leftarrow ((HL))$
MOV M,r: $((HL)) \leftarrow (r)$

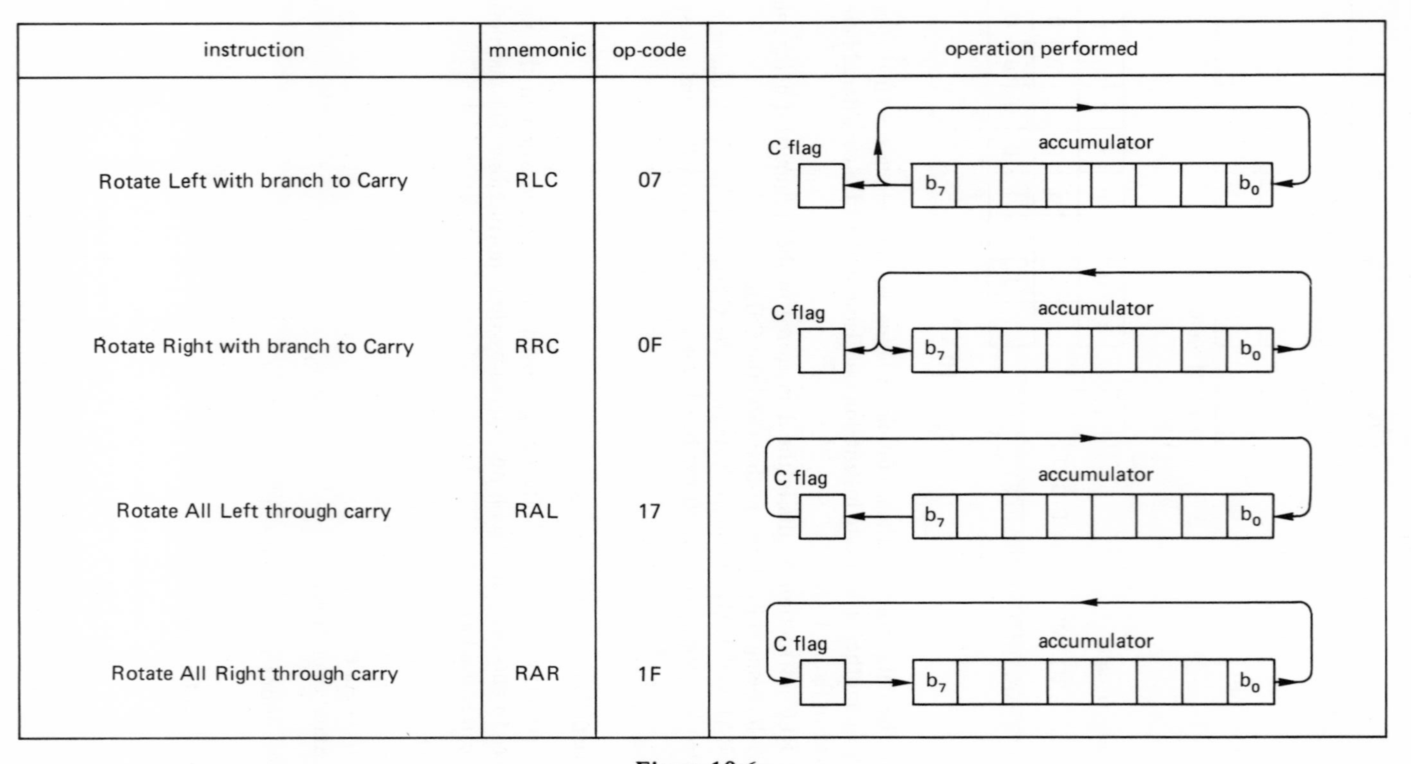

instruction	mnemonic	op-code	operation performed
Rotate Left with branch to Carry	RLC	07	C flag, accumulator b_7 … b_0
Rotate Right with branch to Carry	RRC	0F	C flag, accumulator b_7 … b_0
Rotate All Left through carry	RAL	17	C flag, accumulator b_7 … b_0
Rotate All Right through carry	RAR	1F	C flag, accumulator b_7 … b_0

Figure 10.6

Table 10.9 Data transfer instructions

Least significant hex character (columns); Most significant hex character (rows)

	0	1	2	3	4	5	6	7	8	9	A	B	C	D	E	F
0		LXI B	STAX B				MVI B				LDAX B				MVI C	
1		LXI D	STAX D				MVI D				LDAX D				MVI E	
2		LXI H	SHLD				MVI H				LHLD				MVI L	
3		LXI SP	STA				MVI M				LDA				MVI A	
4	MOV B,B	MOV B,C	MOV B,D	MOV B,E	MOV B,H	MOV B,L	MOV B,M	MOV B,A	MOV C,B	MOV C,C	MOV C,D	MOV C,E	MOV C,H	MOV C,L	MOV C,M	MOV C,A
5	MOV D,B	MOV D,C	MOV D,D	MOV D,E	MOV D,H	MOV D,L	MOV D,M	MOV D,A	MOV E,B	MOV E,C	MOV E,D	MOV E,E	MOV E,H	MOV E,L	MOV E.M	MOV E.A
6	MOV H,B	MOV H,C	MOV H,D	MOV H,E	MOV H,H	MOV H,L	MOV H,M	MOV H,A	MOV L,B	MOV L,C	MOV L,D	MOV L,E	MOV L,H	MOV L.L	MOV L.M	MOV L.A
7	MOV M,B	MOV M,C	MOV M,D	MOV M,E	MOV M,H	MOV M,L		MOV M,A	MOV A,B	MOV A,C	MOV A,D	MOV A,E	MOV A,H	MOV A,L	MOV A,M	MOV A,A
8																
9																
A																
B																
C		POP B				PUSH B										
D		POP D		OUT		PUSH D						IN				
E		POP H		XTHL		PUSH H						XCHG				
F		POP PSW				PUSH PSW				SPHL						

The address of the memory is specified by the contents of the register pair HL, and r is one of the registers A, B, C, D, E. For example, the instruction MOV M,B (op-code 70) results in the contents of register B being copied into the RAM address specified by the contents of the register pair HL.

The 2-byte instruction MVI M,data (op-code 36) causes the data in byte 2 of the instruction to be MoVed Immediately to the Memory location specified by the register pair HL as follows

MVI M,data: ((HL)) ← (byte 2)

A number of instructions are included in the instruction set which allow the programmer to load the accumulator direct from memory (LDA - op-code 3A), or to store the contents of the accumulator directly into memory (STA - op-code 32). Both are 3-byte instructions, the first byte being the op-code, the second byte being the low-order byte of the memory address, and the third byte being the high-order byte of the memory address. These instructions are described by the following.

LDA addr: (A) ← ((byte 3) (byte 2))
STA addr: ((byte 3) (byte 2)) ← (A)

Two 3-byte instructions are available which load and store, respectively, the register pair from two successive memory locations. These instructions are defined below and one of them, the LHLD (Load H and L Direct) instruction is illustrated in figure 10.7.

LHLD addr: (L) ← ((byte 3) (byte 2))
(H) ← ((byte 3) (byte 2) + 1)
SHLD addr: ((byte 3) (byte 2)) ← (L)
((byte 3) (byte 2) + 1) ← (H)

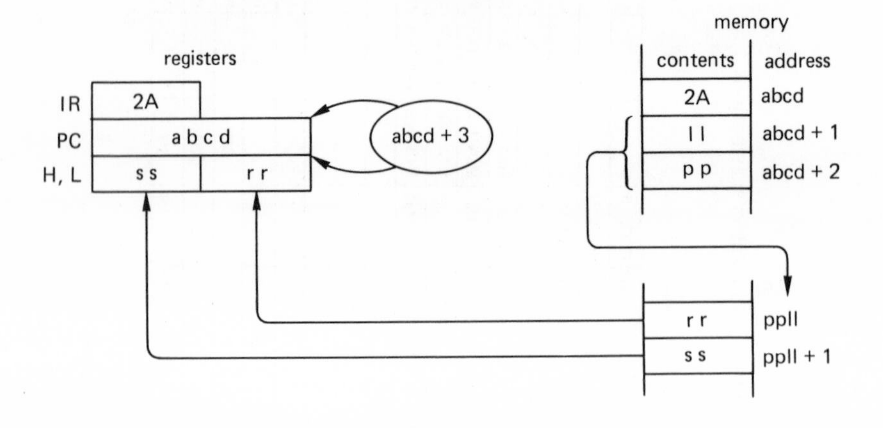

Figure 10.7

Instructions are also available which allow the programmer to load the accumulator or to store the contents of the accumulator using register indirect addressing. The load instructions are

LDAX B: (A) ← ((BC))
LDAX D: (A) ← ((DE))

If register D contains 05_{16}, register E contains 86_{16}, and memory location 0586_{16} contains $1E_{16}$, then after the execution of a LDAX D instruction (op-code 1A) the accumulator contains 1E.

The register indirect instructions which enable the contents of the accumulator to be stored in a memory location are

STAX B: ((BC)) ← (A)
STAX D: ((DE)) ← (A)

In this case, the contents of the accumulator are stored in the address specified either by the register pair DC, or the pair DE.

Register transfer operations

These are internal transfer instructions used to move data between registers, or to load a register (or a register pair). The simplest type of register transfer operation is the 1-byte MOV r1,r2 (MOVe register to register) instruction which uses register direct addressing as follows

MOV r1,r2: (r1) ← (r2)

An example of this instruction is the MOV D,B instruction (op-code 50). This instruction causes the contents of register B to be transferred to register D. Certain of these instructions correspond to a 'do nothing' instruction; an example being the MOV B,B instruction (op-code 40) which results in the contents of register B being moved into register B.

A number of instructions, using immediate addressing, allow any register to be loaded with data; these are generally similar to the MVI M instruction described earlier. Examples of this type of instruction include MVI B, MVI D, MVI H, MVI C, MVI E, MVI L, and MVI A (see table 10.9).

In addition, instructions are included in the instruction set which allow a register pair to be loaded with data; once again, immediate addressing is used. These are the 3-byte LXI rp instructions, the first byte being the op-code (LXI B - 01, LXI D - 11, LXI H - 21, LXI SP - 31), the second byte contains data which is the data moved to the low-order register of the pair and the third byte is the data moved to the high-order register of the pair. An example of the use of the LXI SP instruction is given in table 8.2.

An instruction is also available which exchanges the contents of the register pair HL with the contents of register pair DE. This is the 1-byte XCHG instruction (eXCHanGe HL with DE - op-code EB) which uses register direct addressing, and results in the following operation

$$\text{XCHG:} \quad (H) \leftrightarrow (D), (L) \leftrightarrow (E)$$

I/O data transfers

Included in this group are the INput data from an I/O port (op-code DB) and the OUTput data to an I/O port (op-code D3). These instructions are described in chapter 7.

Stack operations

During interrupt routines and subroutine operations, it may be necessary to save the contents of the registers on the stack. Alternatively, at some other time, it becomes necessary to return the contents from the stack to the registers. The above data movements are initiated by PUSH and POP instructions, respectively, and are described in chapter 8.

Additionally, two 1-byte instructions are available which allow the interchange of the contents of the stack pointer with the contents of the register pair HL. These are XTHL (eXchange stack Top with H and L) and SPHL (move HL to SP) as follows.

$$\text{XTHL (op-code E3):} \ (L) \leftrightarrow ((SP)), (H) \leftrightarrow ((SP) + 1)$$
$$\text{SPHL (op-code F9):} \ (SP) \leftarrow (H)\ (L)$$

Using the SPHL instruction, the user can access two independent stacks.

10.18 Program manipulation instructions

This group of instructions permits the transfer of control from one place in the program to another and includes

(1) Unconditional jump instructions
(2) Conditional jump instructions
(3) Subroutine instructions

The instructions in the above category are listed in table 10.10.

Unconditional jumps

An instruction in this group causes program control to be transferred unconditionally to the address specified in bytes 2 and 3 of the instruction. The instruction is described by the statement

$$\text{JMP addr:} \quad (PC) \leftarrow (\text{byte } 3)\ (\text{byte } 2)$$

Table 10.10 Program manipulation instructions

Rows: Most significant hex character; Columns: Least significant hex character

	0	1	2	3	4	5	6	7	8	9	A	B	C	D	E	F
0																
1									JR							
2	JRNZ								JRZ							
3	JRNC								JRC							
4																
5																
6																
7																
8																
9																
A																
B																
C	RNZ		JNZ	JMP	CNZ			RST 0	RZ	RET	JZ	JMPI	CZ	CALL		RST 1
D	RNC		JNC		CNC			RST 2	RC		JC		CC			RST 3
E	RPO		JPO		CPO			RST 4	RPE	PCHL	JPE		CPE			RST 5
F	RP		JP		CP			RST 6	RM		JM		CM			RST 7

The above instruction (op-code C3) uses immediate addressing; this instruction is illustrated in table 7.1.

The 3-byte indirect addressed unconditional jump instruction JMPI (op-code CB), causes control to be transferred to an address which is specified by the contents of an address in the second and third bytes of the JMPI instruction (see also figure 10.1).

The unconditional jump address can also be specified by program relative addressing in the form of a 2-byte JR instruction (see also table 10.6).

An unconditional jump can also be made to an address specified by the contents of the register pair HL using the 1-byte PCHL (move to PC the contents of register pair HL - op-code E9) which uses register direct addressing as follows

$$\text{PCHL:} \quad (\text{PC higher}) \leftarrow (\text{H}), (\text{PC lower}) \leftarrow (\text{L})$$

After the execution of a PCHL instruction, the PC stores a copy of the register pair HL; control is then transferred to that address.

Conditional jump instructions

A range of conditional jump instructions is available to the programmer. If the specified condition is true (for example, jump to a given address IF the zero status flag is set), control is transferred to the address specified in the instruction. A series of 3-byte conditional jump instructions using immediate addressing is available, the condition for the jump depending on the value stored in the appropriate status flag as follows

JNZ	zero flag not set (Z = 0)
JZ	zero flag set (Z = 1)
JNC	no carry (C = 0)
JC	carry (C = 1)
JPO	parity odd (P = 0)
JPE	parity even (P = 1)
JP	plus (S = 0)
JM	minus (S = 1)

Alternatively, if the address is specified by relative addressing, one of the JR series of instructions (JRNZ, JRNC, JRZ and JRC) is used (see also table 10.6).

Subroutine instructions

These instructions include the 3-byte unconditional subroutine CALL instruction (op-code CD - see figure 8.1 and associated text), the 1-byte unconditional RETurn instruction (op-code C9), and a range of conditional CALL and RETurn instructions. For information about the latter, the reader is referred to tables 8.11 and 8.12.

Subroutines can also be called by a number of RST (ReSTart) instructions which are given in tables 8.9 and 8.10.

10.19 Status Management and other Instructions

Status management instructions are concerned with alteration of the status flags of the CPU without affecting the data stored in the microcomputer. In the CPU described here, these are limited to the following 1-byte instructions

EI (Enable Interrupts)	op-code FB
DI (Disable Interrupts)	op-code F3
STC (SeT Carry)	op-code 37
CMC (CoMplement Carry)	op-code 3F

The EI and DI instructions are described in chapter 9. The STC instruction sets the C (carry) flag, while the CMC instruction causes the contents of the carry flag to be complemented.

Other instructions not described so far are:

HLT (HaLT)	op-code 76
NOP (No OPeration)	op-code 00

The HLT instruction causes the processor to stop operations. The NOP instruction results in the CPU doing nothing for the duration of the instruction; this instruction can be used as a means of providing a small adjustment in a time delay program. Neither the HLT nor the NOP instruction alters the status flags.

10.20 Index Register Instructions

The first byte of the IXR instructions (IndeX Register) - op-code DD - refers to operations with the index register IX. A typical 4-byte instruction in this group is described in table 10.2, and other instructions are listed in table 10.4.

PROBLEMS

10.1 Discuss applications where the following methods of addressing would be used: (a) direct addressing, (b) indirect addressing, (c) immediate addressing, (d) indexed addressing, (e) relative addressing, (f) register direct addressing, (g) register indirect addressing, (h) stack addressing.

10.2 In certain CPUs, direct addressing cannot be used, but relative addressing and indexed addressing are available. With the aid of simple programs, explain how data can be read from and stored in memory locations under program control using such a microprocessor.

11 Programming and Applications

11.1 An Introduction to Programming

A **program** is a sequence of instructions which results in the CPU performing a required task. The sample programs included in this chapter include many instructions and techniques which are frequently used by programmers. In some cases it it possible to shorten the sample programs by using sophisticated techniques.

For all but very simple programs, the reader is advised to draw up an **outline flowchart** which defines the problem. The flowchart

(1) aids the programmer to convert the problem specification to a sequential list of instructions
(2) provides a guide for further development of the program
(3) ensures that the program specification is met

In addition, it may then be advisable to draw up a **detailed flowchart**, which provides sufficient detail to enable the program to be written in the programming language used by the CPU.

Where a complete program is illustrated in this chapter, the starting address is 0000_{16}. The reader should consult the user's manual of the microprocessor he is using, since certain CPUs do not permit the use of this location for data or instructions. Where a particular technique is illustrated which occurs, for example, in the middle of a program, then the starting address of the illustrative program is other than 0000.

Programs can be terminated using any one of several methods. One method is to use a HLT instruction; to 'escape' from a HLT state, it is either necessary to generate a hardware interrupt (see chapter 9) or the computer must be reset (which enables the CPU to return to program address 0000). Alternatively, the program can be terminated by means of a RESTART instruction (which must have a suitable subroutine associated with it). Program termination can also be by means of an unconditional JUMP instruction which transfers control to the monitor program. Yet another method of terminating a program is to cause the CPU to execute a continuous program loop (see also table 11.1).

Method A in table 11.1 uses a HLT instruction to terminate the program.

Table 11.1 Some methods of program termination

Address (hex)	Method A	Method B	Method C	Comment
007F	XX	XX	XX	Final instruction in program being executed
0080	76 HLT	C3 JMP	18 JR	Methods of program termination
0081		80 } 0080	FE -2_{10}	
0082		00 }		

Method B causes the CPU to execute an endless loop, since it causes control to continuously be transferred to location 0080; the loop is broken by resetting the CPU. Method C is similar to method B in that the program executes a continuous loop, but this time by means of a program counter relative jump of -2_{10}. Once again, the CPU executes an endless loop until it is reset.

In the case of a microcomputer development system, one instruction can be used to cause a 'return' to monitor' jump. This instruction can be used to terminate the program. For example, in the Acorn microcomputer development system (which uses a 6502 CPU), the CPU returns control to the monitor when it receives the 3-byte instruction 4C 04 FF (unconditional jump to address FF04).

11.2 8-bit Addition

This program (see table 11.2) adds together the contents of locations 0090 and 0091, and stores the result in location 0092. The programmer must ensure that the result of the addition does not exceed FF_{16}, otherwise an overflow is generated (this situation is covered in section 11.3).

Table 11.2 8-bit addition

Address	Machine code	Mnemonic	Comment
0000	21	LXI H,0090_{16}	; Load HL with 0090
01	90		
02	00		
03	7E	MOV A,M	; (A) ← (0090) first operand
04	23	INX H	; (HL) = 0091
05	86	ADD M	; (A) ← (A) + ((HL)) add second operand
06	23	INX H	; (HL) = 0092
07	77	MOV M,A	; (0092) ← (A) store result
0008	76	HLT	; halt

The LXI 0090 instruction causes register pair HL to store address 0090; the instruction MOV A,M causes the contents of address 0090 to be transferred to the accumulator. Next, the register pair HL is incremented so that it stores address 0091. Instruction ADD M results in the contents of address 0091 being added to the contents of the accumulator, the result being retained in the accumulator. That is

$$(A) = (0090) + (0091)$$

The contents of the register pair HL are incremented once more so that they store address 0092. Finally, the MOV M,A instruction results in the sum being copied into address 0092. If $(0090) = 16_{16}$ and $(0091) = 0F_{16}$, then after the completion of the program $(0092) = 25_{16}$.

11.3 16-bit Addition

Microprocessors having an instruction set of the type described in chapter 10, can perform double-precision addition or double-length addition with the same ease as 8-bit addition. In the following, each 16-bit number is stored in the form of two 8-bit values in adjacent address locations. Suppose that the first 16-bit operand is stored in locations 0090 and 0091 and the second 16-bit operand in locations 0092 and 0093, with the most significant byte of the numbers being stored in locations 0091 and 0093, respectively. The result is to be stored in locations 0094 (l.s.b.) and 0095 (m.s.b).

A program for 16-bit addition is listed in table 11.3. In this case 16-bit data manipulation instructions are used; these reduce the length of the program when

Table 11.3 16-bit addition

Address	Machine code	Mnemonic	Comment
0000	2A	LHLD 0090_{16}	; load HL from 0091 and 0090
01	90		
02	00		
03	EB	XCHG	; transfer (HL) to (DE)
04	2A	LHLD 0092_{16}	; load HL from 0093 and 0092
05	92		
06	00		
07	19	DAD D	; $(HL) \leftarrow (HL) + (DE)$
08	22	SHLD 0094_{16}	; store (HL) in 0095 and 0094
09	94		
0A	00		
0B	76	HLT	; halt

compared with the case where 8-bit instructions are used to perform a 16-bit addition.

11.4 8-bit Subtraction

In this case we illustrate subtraction using register addressing, the program used being given in table 11.4 and results in the data in register B being subtracted from the data in register A (the accumulator), the result being saved in register C.

Table 11.4 8-bit subtraction

Address	Machine code	Mnemonic	Comment
0000	3E	MVI A,$5A_{16}$	; load minuend into register A
01	5A		
02	06	MVI B,23_{16}	; load subtrahend into register B
03	23		
04	90	SUB B	; subtract (B) from (A) and leave the result in A
05	4F	MOV C,A	; move the result into register C
06	76	HLT	; halt

11.5 Bit Testing

In many applications it is necessary to test the state of one bit in the accumulator to see if it is a '0' or a '1'. The two most popular methods used in this application are

(1) Using the logical AND instruction together with a 'word mask', followed by a conditional jump instruction
(2) Using a rotate instruction and the carry flag to test the selected bit

Both methods are described below. To illustrate the need for bit testing, consider the case of an elevator (lift) control system. The position of the elevator is detected by a microswitch at each level; when the elevator reaches one of the floors, it causes the microswitch at that level to produce a logic '1', otherwise its output is logic '0'. Suppose that there are eight floors (including the ground floor [floor zero]), and that the microswitch on each floor is connected to the microcomputer data bus via an I/O port whose address is 01 (see also chapter 7). In each of the following programs, tests are made to determine when the lift has reached the third floor.

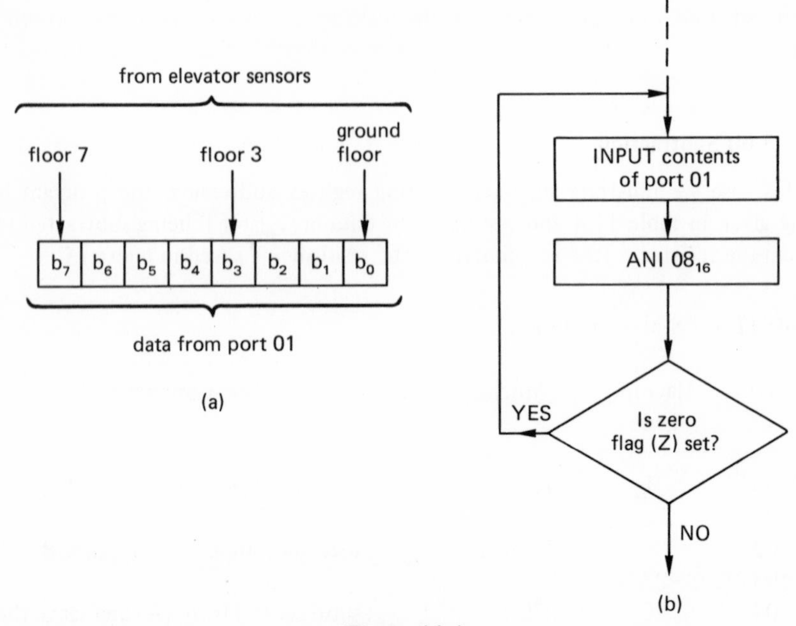

Figure 11.1

Masking technique using the ANI instruction

A flowchart illustrating the general procedure is shown in figure 11.1. The logic signal from the microswitch on each floor is connected to the accumulator via input port 01, with the state of the ground floor switch being transferred to bit b_0 in the accumulator, the first floor switch to b_1, the third floor to b_3, etc. (see figure 11.1a). The flowchart initially calls for data from port 01 to be transferred

Table 11.5 Bit testing using the ANI instruction

Address	Machine code	Label	Mnemonic	Comment
0051	XX			; part of main program
52	DB	INPORT:	IN 01	; input data from port 01
53	01			
54	E6		ANI 08_{16}	; mask out all bits except b_3
55	08			
56	CA		JZ INPORT	; jump if (A) zero
57	52			
58	00			
59	XX			; continue if (A) non-zero

to the accumulator. Following this, the contents of the accumulator are ANDed with 08_{16} ($0000\,1000_2$); when the elevator is not at the third floor, the result of the ANDing operation is zero (that is, the zero (Z) flag is set). At this stage, the elevator has not reached its destination, and the flowchart indicates that the CPU must repeat its test on the state of port 01 once more. The waiting loop is continuously cycled by the CPU until the elevator arrives at the third floor; when it does, the result of ANDing the accumulator contents with 08_{16} is non-zero. When this condition is detected, the CPU emerges from the 'NO' exit of the decision symbol and can continue with the remainder of the program (which would be to stop the lift drive and open the doors.) The corresponding program is given in table 11.5.

The 3-byte JZ instruction at location 0056 in table 11.5 can be replaced by the 2-byte program counter relative jump instruction JRZ as follows

Address	Machine code	Mnemonic	Comment
0056	20	JRZ INPORT	;jump relative to 0058
57	FA		;by -6_{10}
58	XX		; continue if (A) non-zero

Bit testing using rotate instructions and testing the carry flag

In this case the contents of the accumulator are rotated until a copy of the required bit appears in the carry flag. The carry flag is then tested to determine its value; if it is zero, then port 01 is again sampled and the sequence repeated. If the carry flag is set as a result of the rotate instructions, the CPU is allowed to continue with the remainder of the operation. A program illustrating this technique is illustrated in table 11.6. Referring to the table the reader will see that, after transferring data from port 01, the contents of the accumulator are rotated right with branch to carry four times. The final RRC instruction causes the data originally in the b_3 position in the accumulator to be copied into the carry flag. If the carry flag is then zero the JNC jump is made to the label INPORT in the program, otherwise the instruction at location 005B is executed.

The 3-byte immediate addressing jump instruction JNC can be replaced by a 2-byte program counter relative jump instruction as follows

Address	Machine code	Mnemonic	Comment
0058	30	JRNC INPORT	;jump relative to 005A
59	F8		;by -8_{10}
5A	XX		; continue if C flag non-zero

Table 11.6 Bit testing using rotate instructions

Address	Machine code	Label	Mnemonic	Comment
0051	XX			; part of main program
52	DB	INPORT:	IN 01	; input data from port 01
53	01			
54	0F		RRC	; rotate right with branch
55	0F		RRC	; to carry, four times
56	0F		RRC	
57	0F		RRC	
58	D2		JNC INPORT	; jump to INPORT if carry
59	52			; flag not set
5A	00			
5B	XX			; continue if C flag not set

Test for logic '0'

If, in the programs listed in tables 11.5 and 11.6, we need to test for logic '0' rather than logic '1', it is merely necessary in table 11.5 to alter instruction 0056 to JNZ (op-code C2), and in table 11.6 to alter instruction 0058 to JC (op-code DA).

11.6 8-bit x 8-bit Multiplication

The process of binary multiplication was described in section 2.18. It can be represented by a number of programs, one of which is given in table 11.7. Here the multiplicand (Md) is stored in the C register the multiplier (Mr) is stored in the D register, and the product is stored in the register pair HL. The associated flowchart is given in figure 11.2.

The operations are carried out on register pairs, the Md is stored in the least-significant byte of register pair BC, and the Mr is stored in the most-significant byte of the register pair DE. Since two 8-bit numbers are to be multiplied, the calculation is complete after eight shift-and-add sequences. A counter (the accumulator) is loaded with 08_{16} (corresponding to the number of shift-and-add operations to be carried out); the value stored in the counter is decremented after each shift-and-add sequence, so that when the counter stores zero the program is terminated. The initialisation of the program is completed after the execution of the LXI H,0000_{16} instruction, which clears the result register.

The sequence of instructions XCHG, DAD H and XCHG (instructions 000B to 000D, inclusive) shifts the m.s.b. of the multiplier into the carry flag while preserving the value stored in the product register. The reader will note that instruction 000C (that is, DAD H) is described as shifting the contents of the register pair

Table 11.7 8-bit x 8-bit multiplication

Address	Machine code	Label	Mnemonic	Comment
0000	01		LXI B,0006_{16}	; Md = 06
01	06			
02	00			
03	11		LXI D,0500_{16}	; Mr = 05
04	00			
05	05			
06	3E		MVI A,08_{16}	; count = 08
07	08			
08	21		LXI H,0000_{16}	; product = 0000
09	00			
0A	00			; initialisation complete
0B	EB	NEXT:	XCHG	; (HL) ↔ (DE)
0C	29		DAD H	; shift (HL) left
0D	EB		XCHG	; (HL) ↔ (DE)
0E	D2		JNC NOCNT	; jump if m.s.b. of Mr is '0'
0F	12			
10	00			
11	09		DAD B	; product = product + Md
12	3D	NOCNT:	DCR A	; have 8 steps been completed?
13	CA		JZ END	; yes, finish
14	1A			
15	00			
16	29		DAD H	; no, shift product one bit left
17	C3		JMP NEXT	; repeat routine
18	0B			
19	00			
1A	76	END:	HLT	; halt

HL one place left. It appears unusual to use an addition instruction to cause a shift left, its use being explained in the following. When a number is added to itself, the result is double the original value; when this operation is carried out on a binary number, each bit in the number is shifted one place left. The simplest method of shifting a 16-bit number one place left is to cause the CPU to execute a double-length addition (see instruction 000C in table 11.7). If the carry flag is set after these instructions, the multiplicand is added to the product; if it is '0', the product is unaltered.

The DCR A instruction decrements the contents of the accumulator (the counter), and a check is then made to see if the counter stores zero. If it does, the program is terminated by jumping to the label END; if not, the CPU executes a DAD H instruction which results in a one-bit shift left of the register pair HL. The whole process is repeated until the counter stores zero.

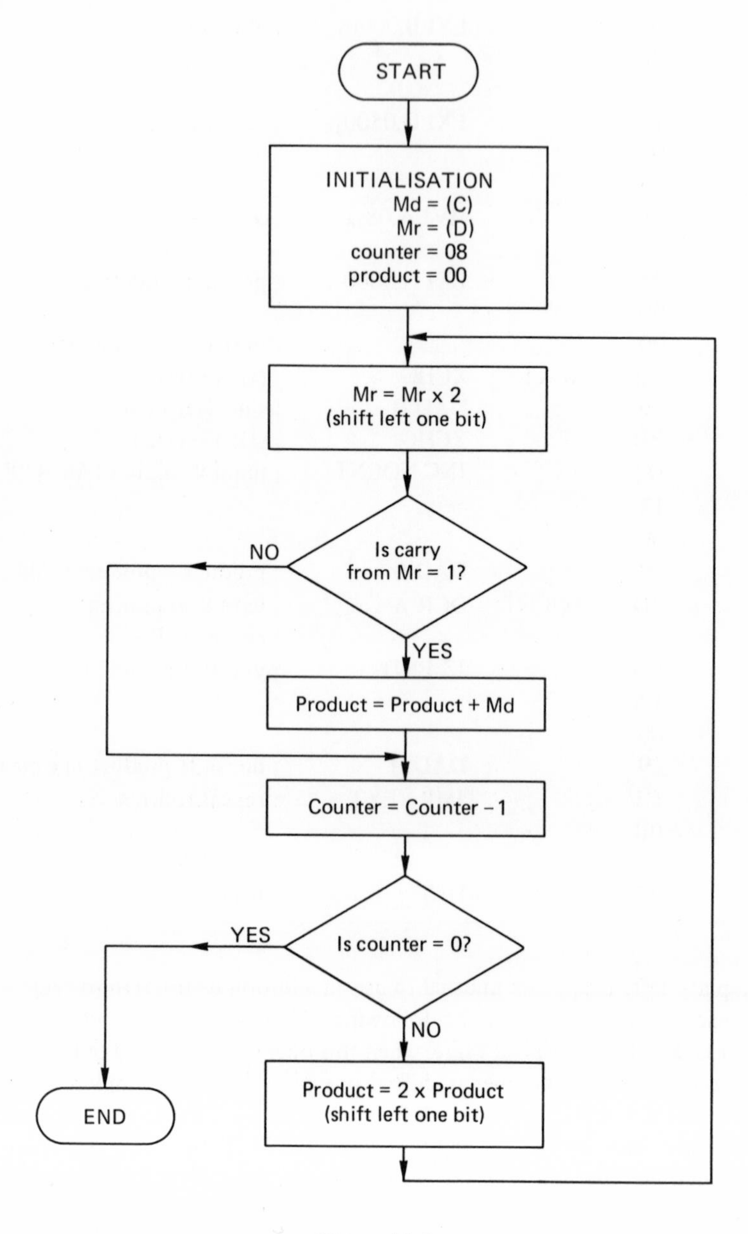

Figure 11.2

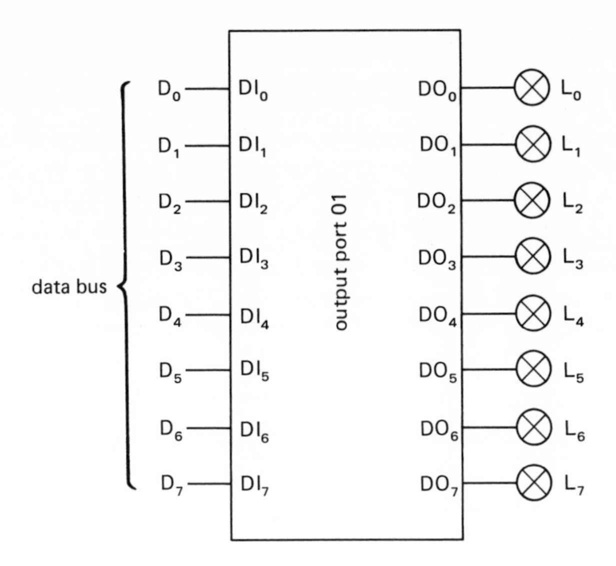

Figure 11.3

11.7 Flashing Light Sequence

Let us suppose for the moment that we have available a microcomputer which has an output port whose address is 01, and each line of the port is connected to a lamp. The connections between the output port and the lamps are as shown in figure 11.3 (the chip select lines are not shown - for details see figure 5.1).

The program developed below results in the lights being illuminated in the sequence in table 11.8, in which a '1' corresponds to an illuminated lamp and '0' to an extinguished lamp.

Initially, all the lamps remain extinguished for a period of 0.2 s (corresponding to the 'initial conditions' in table 11.8). After this, lamps L_7 and L_6 must be illuminated for 0.2 s, after which they are extinguished and L_5 and L_4 are illuminated for 0.2 s. This procedure is repeated until all the lamps have been illuminated in pairs for 0.2-s intervals. Following the period when L_1 and L_0 are

Table 11.8 Lamp control sequence

Step	L_7	L_6	L_5	L_4	L_3	L_2	L_1	L_0	
Initial condition	0	0	0	0	0	0	0	0	←
1	1	1	0	0	0	0	0	0	
2	0	0	1	1	0	0	0	0	Return
3	0	0	0	0	1	1	0	0	
4	0	0	0	0	0	0	1	1	

illuminated, all the lamps must be extinguished for 0.2 s once more, and the cycle recommenced.

A flowchart giving one solution to the problem is provided in figure 11.4, the corresponding program is given in table 11.9. Referring to the program, the reader will see that 00_{16} is output to port 01, causing all the lamps to be extinguished. Next, a time delay subroutine, TIME 1, is CALLed by the program. This subroutine provides a 0.2-s software time delay, and is described in chapter 8; the flowchart of the time delay program is given in table 8.6 and its program in table 8.6, the program commencing at address 0110_{16}. The effect of TIME 1 at this point in the program is to ensure that all the lamps remain extinguished for 0.2 s.

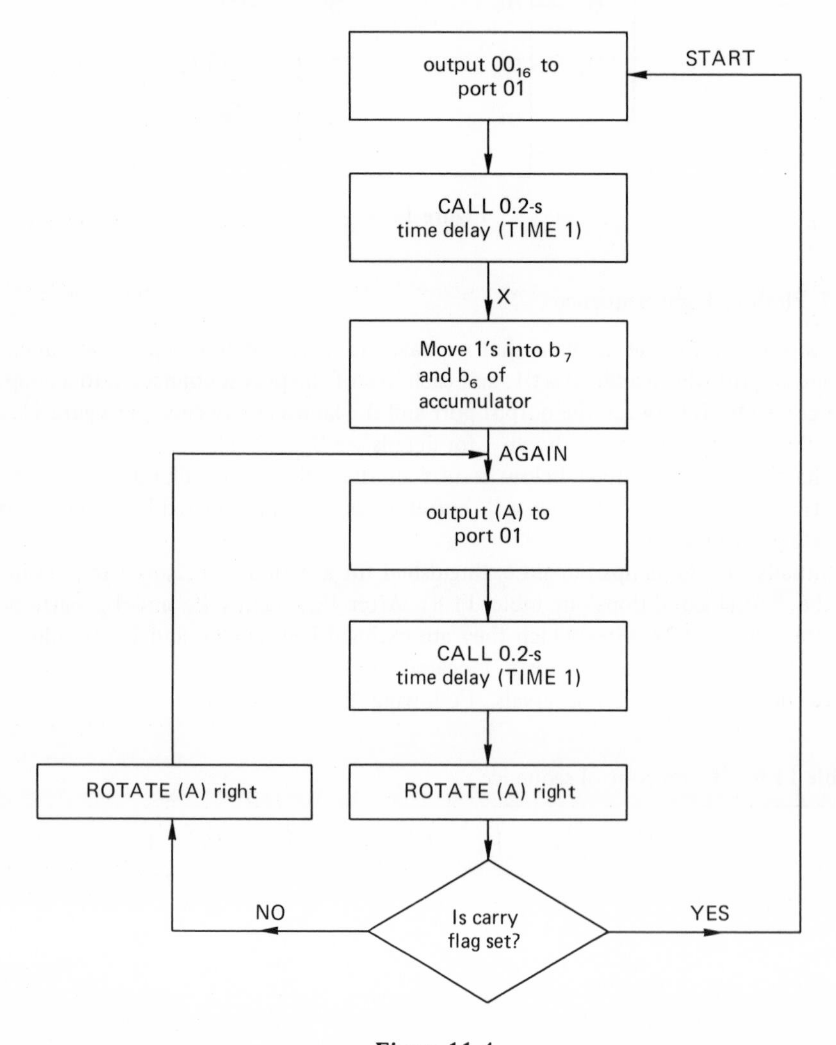

Figure 11.4

Table 11.9 A program for a flashing light sequence

Address	Machine code	Label	Mnemonic	Comment
0000	3E	START:	MVI A,00_{16}	; (A) ← 0000 0000_2
01	00			
02	D3		OUT 01	; extinguish all lights
03	01			
04	CD		CALL TIME 1	; maintain lights out
05	10			; for 0.2s
06	01			
07	00		NOP	; required for program
08	00		NOP	; modification later
09	00		NOP	
0A	3E		MVI A,1100 0000_2	
0B	CO			
0C	D3	AGAIN:	OUT 01	; illuminate two lamps
0D	01			
0E	CD		CALL TIME 1	; maintain two lamps
0F	10			; on for 0.2s
10	01			
11	0F		RRC	; rotate (A) right
12	DA		JC START	; check for end of sequence
13	00			
14	00			
15	0F		RRC	; continue with sequence
16	C3		JMP AGAIN	; illuminate two other lamps
17	0C			
18	00			

Addresses 0007, 0008 and 0009 are each filled with NOP instructions (No OPeration - op-code 00). These locations are to be filled later with other instructions, and the effect of the NOPs on the program at this stage is merely to increase the time delay by a minute amount of time. These instructions correspond to the point marked X on the flowchart in figure 11.4.

The pair of instructions MVI A,1100 0000_2 and OUT 01, result in logic 1's being applied to L_7 and L_6, and logic 0's being applied to the remainder of the lamps. The subroutine TIME 1 is CALLed for once again to illuminate L_7 and L_6 for 0.2 s.

After this period of time, it is necessary to check if the lamp-flashing sequence has been completed. The method adopted here is as follows (see also figure 11.4 and table 11.9). The CPU executes an RRC instruction (Rotate accumulator contents Right with branch to Carry), which results in the state of the l.s.b. of the accumulator being copied into the carry flag. If, after the first RRC instruction is executed, the carry flag is not set then the flashing sequence is not complete. In this case the CPU executes the second RRC instruction; the accumulator then contains $0011\ 0000_2$.

The program then returns to the point labelled AGAIN, when the contents of the accumulator are output to the lamps for 0.2 s. This results in lamps L_7 and L_6 being extinguished, and L_5 and L_4 illuminated.

The 'AGAIN' loop is cycled several times more until lamps L_1 and L_0 are illuminated for 0.2 s. After this, the first RRC instruction in the next 'AGAIN' loop results in the carry flag being set (this occurs because the logic '1' in bit b_0 of the accumulator is 'rotated' into the C flag). Consequently, an exit is made from the YES exit of the 'Is the Carry Flag set?: decision symbol. This results in an orderly return to the START of the program. The lamps are then extinguished again prior to the start of the flashing lamp sequence.

11.8 Switch-controlled Flashing Light Sequence

The sequence of flashing lights described in section 11.7 continues idefinitely so long as the power supply to the CPU is maintained. Let us suppose, however, that the sequence of events in table 11.8 corresponds, say, to a number of operations carried out on a single component on a production line. In this case, the program must be modified so that the CPU waits for a component to present itself before commencing the sequence.

For the purpose of discussion, we will assume that the program must halt at point X (see the flowchart in figure 11.4) when waiting for a component on the production line.

To prevent the CPU from progressing beyond point X in the program, a waiting loop subroutine (labelled WAIT - see the program in table 11.10) is called at point X. The program in table 11.10 is now described. After the main program outputs 00_{16} to the lamps, it enters the WAITing loop beginning at address 0020. The subroutine causes the data from input port 01 to be input to the accumulator. The sensor which detects the presence of a component on the production line is connected to bit b_0 of this input port, so that when a component presents itself to the sensor, a logic '1' is transferred to bit b_0 of the accumulator (via the input port). The absence of a component is indicated by a logic '0' in bit b_0.

The ANI 01_{16} instruction (AND Immediate the contents of the accumulator with 00000001_2) masks out bits b_7 - b_1 of the accumulator, leaving bit b_0 with either a '1' or a '0', depending on the signal from the sensor on the production line. If $b_0 = 0$ (that is, no component on the production line), control is transferred

Table 11.10 Waiting loop subroutine for lamp sequence

Address	Machine code	Label	Mnemonic	Comment
0000	3E			; start of main program
.				
.				
0007	CD		CALL WAIT	; call waiting loop subroutine
08	20			
09	00			
.				
.				
0020	DB	WAIT:	IN 01	; input contents of port 01
21	01			
22	E6		ANI 01_{16}	; check l.s.b. of port 01
23	01			
24	CA		JZ WAIT	; jump if l.s.b. of port
25	20			; 01 is logic '0'
26	00			
27	C9		RET	; return if l.s.b. is non-zero

to address 0020 once more. The waiting loop is cycled until $b_0 = 1$, when a RETurn is made to the main program at address 000A.

Alternatively, the waiting loop described above could be written in the main program. However, the method used here shows how a program written for one application can be easily modified to deal with an alternative application.

11.9 Simple Traffic Light Control Program

At this point in the book we consider the use of a microcomputer as a dedicated controller in a traffic light scheme. The CPU must monitor the flow of traffic in two directions, and must control the flow of traffic by means of two sets of traffic lights. A plan view of the road junction is given in figure 11.5. The flow of traffic in direction A is detected by sensor S_A, and in direction B by sensor S_B. The traffic flow in direction A is controlled by three lamps, namely a green lamp G_A, an amber lamp AM_A and a red lamp R_A; the traffic flow in direction B is controlled by lamps G_B, AM_B and R_B.

The traffic light sequence is given in table 11.11. At the instant of switch-on the red light is displayed in direction A, halting the flow of traffic in that direction. At the same time, the green light is displayed in direction B, allowing traffic to flow in that direction.

Table 11.11 Traffic light sequence

Direction A		Direction B	
Lamp illuminated	Action	Lamp illuminated	Action
R_A		G_B	
	Test for vehicle in direction A. Traffic flow meanwhile halted in direction A.		
R_A and AM_A		AM_B	
G_A		R_B	
			Test for vehicle in direction B. Traffic flow meanwhile halted in direction B
AM_A		R_B and AM_B	

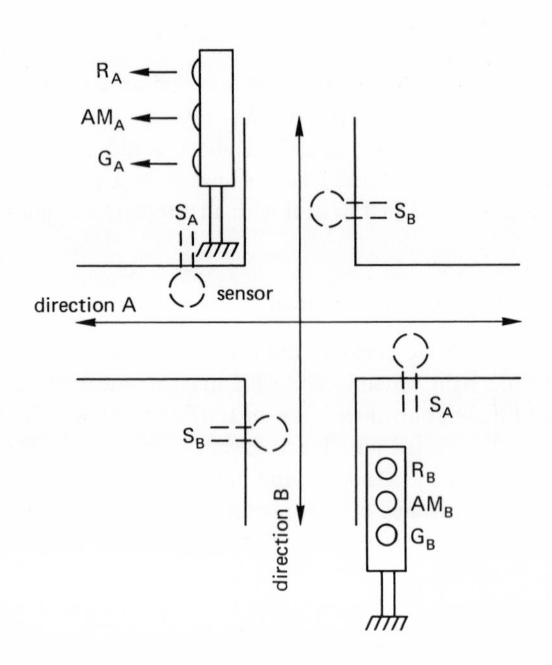

Figure 11.5

When the lamps are in the above state, the CPU scans the condition of sensor A to detect when a vehicle is approaching the lights in direction A. In the following, it is assumed that a vehicle is detected when $S_A = 1$ (conversely, the absence of a vehicle is detected when $S_A = 0$). When a vehicle is detected in direction A, the lamps illuminated must change in the sequence given in table 11.11. The lamp changing sequence is completed when lamps G_A and R_B are illuminated, permitting traffic flow in direction A and halting it in direction B. The CPU must then scan the state of sensor B to detect when a vehicle approaches the lights in direction B.

Since the system illuminates six lamps (note, these are duplicated on opposite sides of the road junction to allow for both directions of traffic flow), only six lines of the output port are needed, that is, one for each colour of the lamp (R_A, AM_A, G_A, R_B, AM_B, G_B). Additionally, the CPU must monitor two traffic sensors (S_A and S_B). Since the total I/O requirement of the system is eight lines (two input lines and six output lines), then an 8-bit programmable I/O port satisfies the requirement. The basic I/O connections to the programmable I/O port are shown in figure 11.6a, with $b_0 - b_2$ being used as output lines supplying signals to the lamps in direction B, $b_3 - b_5$ energising the lamps in direction A, and b_6 and b_7 being input lines from S_B and S_A, respectively. A simplified circuit for one of the traffic sensors is shown in figure 11.6b.

A flowchart for the program is shown in figure 11.7. It is necessary at the outset to define the operation of the PIO; that is to say, bit lines $b_0 - b_5$ must be defined as output lines, and bit lines b_6 and b_7 as input lines (see also sections 7.7 and 7.8 which refer to the operation of the PIO). When the PIO has been defined, lamps R_A and G_B must be illuminated. The CPU then enters a loop in which it 'waits' until a vehicle approaches in direction A, that is, it waits until $S_A = 1$. When a vehicle is

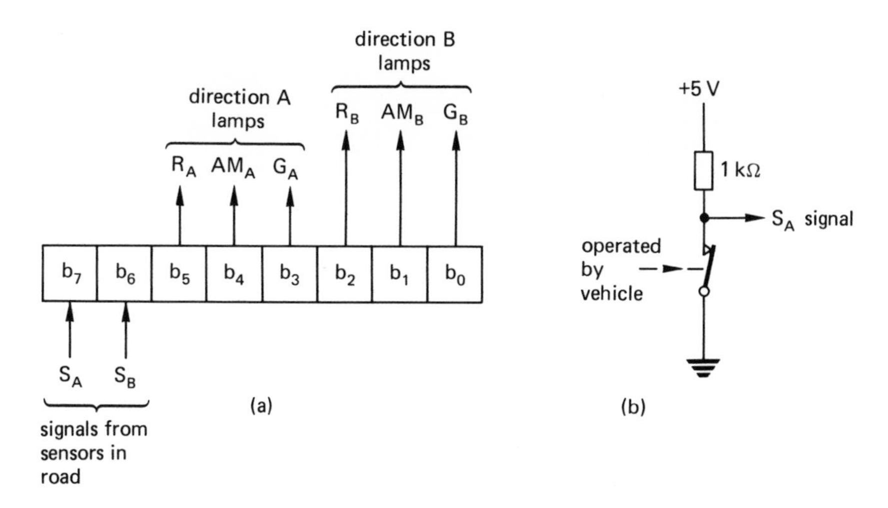

Figure 11.6

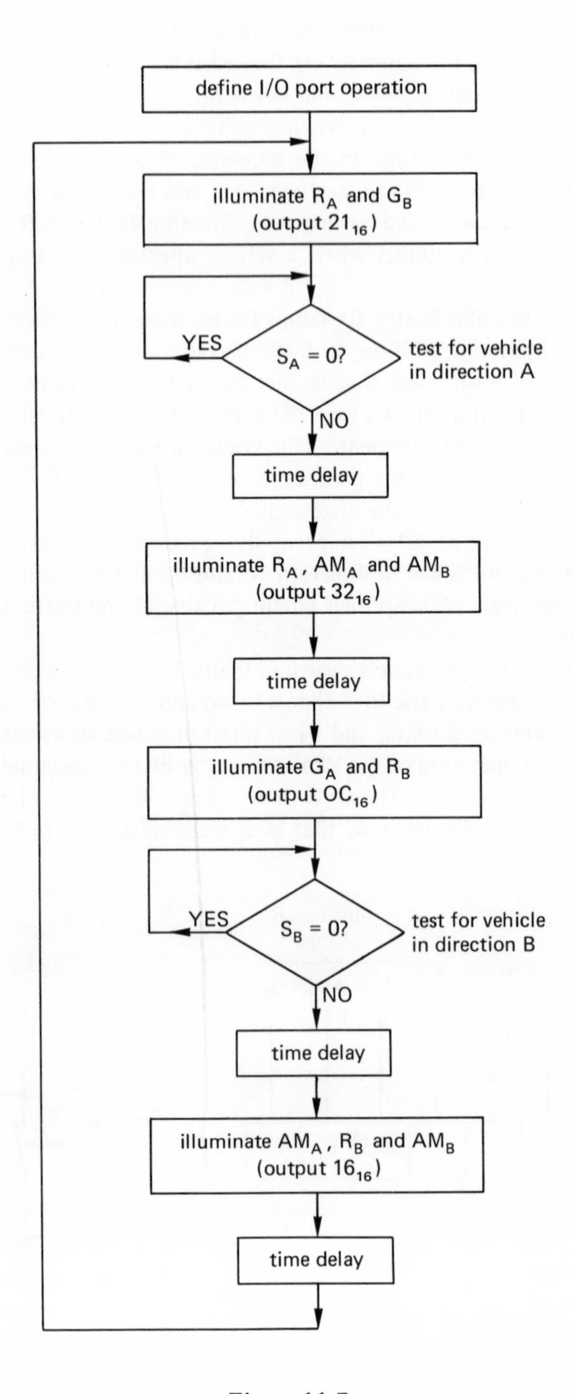
define I/O port operation
illuminate R_A and G_B
(output 21_{16})
YES
$S_A = 0$?
test for vehicle
in direction A
NO
time delay
illuminate R_A, AM_A and AM_B
(output 32_{16})
time delay
illuminate G_A and R_B
(output OC_{16})
YES
$S_B = 0$?
test for vehicle
in direction B
NO
time delay
illuminate AM_A, R_B and AM_B
(output 16_{16})
time delay

Figure 11.7

detected, the CPU executes a time delay program during which time R_A and G_B are illuminated. After this, the lamps change in the sequence listed in table 11.11 until, finally, lamps G_A and R_B are illuminated.

The CPU then waits until a vehicle approaches the junction in direction B; when this occurs, the CPU goes through a second lamp display sequence, at the end of which lamps R_A and G_B are illuminated.

Since a time delay is needed between each set of lamp displays, the time delay sequence is stored in the form of a subroutine. This can be stored as a nested subroutine of the type described in table 8.8; if the time delay between each set of lamp displays is ten seconds, then using the TIME 2 nested subroutine in table 8.8, it is necessary to load register E with 32_{16} (or 50_{10}). The TIME 2 subroutine calls for TIME 1 (which gives a 0.2-s time delay) fifty times, to give a total time delay of ten seconds. If it is necessary to provide differing time delays between each set of lamp displays, then it is merely necessary to modify the program to alter the value moved into register E at the commencement of the TIME 2 subroutine. To simplify the presentation of the traffic light program, the commencing address of TIME 2 is given as $XXYY_{16}$.

The program is given in table 11.12. It is assumed that we are using port A of the PIO port described in chapter 7, and it has the address given in section 7.8. That is to say, the binary word stored in location 0E22 defines the operation of the bit lines

Table 11.12 Simple traffic light sequence program

Address	Machine code	Label	Mnemonic	Comment
0000	3E		MVI A,$3F_{16}$	
01	3F			
02	32		STA 0E22	; define port A
03	22			
04	0E			
05	3E	START:	MVI A,21_{16}	
06	21			
07	32		STA 0E20	; illuminate R_A and G_B
08	20			
09	0E			
0A	3A	TEST A:	LDA 0E20	
0B	20			
0C	0E			
0D	E6		ANI 80_{16}	; test state of S_A
0E	80			
0F	CA		JZ TEST A	; jump if $S_A = 0$
10	0A			
11	00			

Table 11.12 (continued)

Address	Machine code	Label	Mnemonic	Comment
12	CD		CALL TIME 2	; if S_A = 1 CALL TIME 2
13	YY			
14	XX			
15	3E		MVI A,32_{16}	
16	32			
17	32		STA 0E20	; illuminate R_A, AM_A, AM_B
18	20			
19	0E			
1A	CD		CALL TIME 2	; time delay
1B	YY			
1C	XX			
1D	3E		MVI A,$0C_{16}$	
1E	0C			
1F	32		STA 0E20	; illuminate G_A, R_B
20	20			
21	0E			
22	3A	TEST B:	LDA 0E20	
23	20			
24	0E			
25	E6		ANI 40_{16}	; test state of S_B
26	40			
27	CA		JZ TEST B	; jump if S_B = 0
28	22			
29	00			
2A	CD		CALL TIME 2	; if S_B = 1 CALL TIME 2
2B	YY			
2C	XX			
2D	3E		MVI A,16_{16}	
2E	16			
2F	32		STA 0E20	; illuminate AM_A, R_B, AM_B
30	20			
31	0E			
32	CD		CALL TIME 2	; time delay
33	YY			
34	XX			
35	C3		JMP START	; return to START of
36	05			; sequence
37	00			

in port A (a '0' defining an input line, and a '1' defining an output line), the effective address of port A being 0E20. To define b_7 and b_6 as input lines, 0's must be stored in b_7 and b_6 of location 0E22; to define $b_5 - b_0$ as output lines, 1's must be stored in $b_5 - b_0$ of location 0E22. Thus, the first step in the program is to store 00111111_2 ($3F_{16}$) in location 0E22.

The program in table 11.12 should be studied in association with the flowchart in figure 11.7. The binary words which must be output by the CPU to illuminate the various light patterns are given in table 11.13 (see also figure 11.7).

Table 11.13

Lamps illuminated	Binary word to illuminate lamps	Hex word to illuminate lamps
R_A, G_B	0010 0001	21
R_A, AM_A, AM_B	0011 0010	32
G_A, R_B	0000 1100	0C
AM_A, R_B, AM_B	0001 0110	16

At the START of the program proper (address 0005), the CPU outputs 21_{16} to illuminate lamps R_A and G_B. It then enters a waiting loop until a vehicle approaches in direction A. When a vehicle is detected by S_A, the CPU waits for 10 s and then outputs 32_{16} (causing R_A, AM_A and AM_B to be illuminated) and, after a further 10 s, outputs OC_{16} (illuminating G_A and R_B). The CPU then enters a waiting loop once more, but this time it 'looks' for a vehicle approaching in direction B. When a vehicle is sensed, the CPU waits for 10 s and then outputs 16_{16} (causing AM_A, R_B and AM_B to be illuminated) and after a further time delay, it returns to the START of the program.

The above program has two primary drawbacks. Firstly, both sets of traffic lights simultaneously display amber lights; this is potentially dangerous, since it encourages vehicles approaching in both directions to cross the lights. To overcome this disadvantage, the light pattern sequence needs to be altered by re-writing the program; a practical form of light pattern is shown in table 11.14. The second drawback is that during the period of time when the traffic lights are changing from one direction of flow to another, no test is carried out for vehicles approaching the lights. Consequently, a vehicle arriving when the lights are changing from amber to red in this direction will be left stranded at the lights until another vehicle crosses the sensor in his direction of travel. A possible solution to this dilemma is to allow the sensor concerned to generate an interrupt signal during the period when the lights are changing. The interrupt program must, after a suitable time delay, allow the lights to change once more in favour of the stationary vehicle.

Table 11.14

Direction A	Direction B
R	G
test for vehicle	
R	AM
R	R
R and AM	R
G	R
	test for vehicle
AM	R
R	R
R	R and AM

11.10 Digital-to-analog Convertors (DACs)

A DAC is a device (usually a monolithic IC) which converts a binary word at the output of a digital system into an analog signal. A wide variety of DACs exist, and typical word lengths include 8-, 10-, 12- and 14-bits. A wide variety of binary codes are used with these devices including 2's complement, offset binary and binary coded decimal (BCD). DACs using BCD codes are wisely used as interfaces between microcomputers and analog systems. The output voltage range obtainable from a DAC depends on its design, typical ranges being 0–5 V, 0–10 V, ±5 V and ±10 V.

A typical R-2R DAC using only two resistor values, namely R and $2R$, is shown in figure 11.8a. The circuit shown handles a pure binary input word; that is, a logic

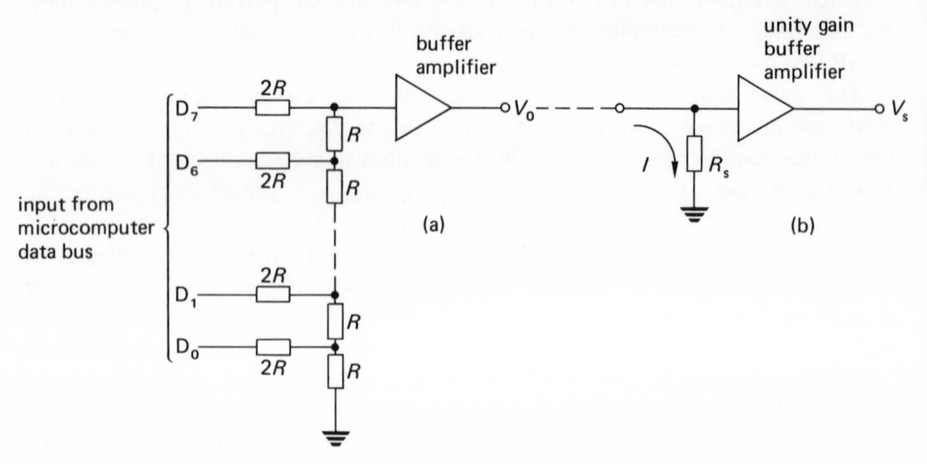

Figure 11.8

'1' on data line D_0 has a decimal 'weight' of 1_{10}, a logic '1' on D_1 has a weight of 2_{10}, the weight of a logic '1' on D_3 is 4_{10}, etc., and the weight of a logic '1' on D_7 is 128_{10}.

In the case of figure 11.8a, the 8-bit data word provides $2^8 = 256$ possible values, that is, 0 to 255. Suppose for the moment that our DAC uses an 8-bit input word, and it converts the input word into an analog output voltage in the range 0 - 1 V. The output voltage change corresponding to a change from '0' to '1' in the l.s.b. of the word is 1/256 V = 0.00390625 V. The output voltage for various binary (and hex) input words are listed in table 11.15.

Table 11.15 DAC output voltage table

Binary input	Hex input	Analog output voltage (V)
0000 0000	00	0.00000000
0000 0001	01	0.00390625
0000 0010	02	0.00781250
.		
.		
0100 0000	40	0.25
0100 0001	41	0.25390625
.		
.		
1111 1110	FE	0.99218750
1111 1111	FF	0.99609375

Some DACs provide an output current rather than an output voltage. If the user wishes to convert the output current into a voltage, then the circuit in figure 11.8b may be used in association with the DAC. The value of R_S is chosen so that when the maximum current flows from the DAC output, it produces the desired maximum value of output voltage V_S.

The DAC in figure 11.8 does not include a data latching circuit. A simple non-latched DAC can be interfaced to the CPU data bus by means of a latched I/O

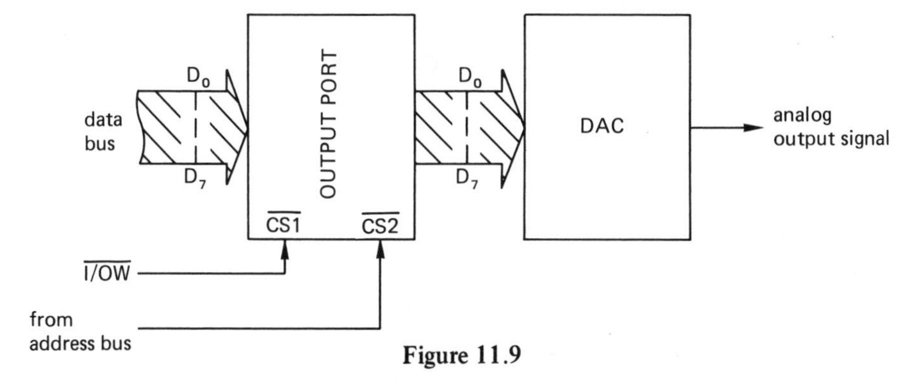

Figure 11.9

port in the manner shown in figure 11.9. The address of the DAC is then the same as that of the I/O port.

11.11 Waveform Generation using a DAC

In the following we consider how simple waveforms can be generated using a DAC in association with a microcomputer, the general circuit being shown in figure 11.9.

Square wave generator

Suppose that we need to generate a square wave which changes from 0 V to 1 V at a frequency of 2.5 Hz (that is, it has periodic time of 0.4 s). The waveform is shown in figure 11.10.

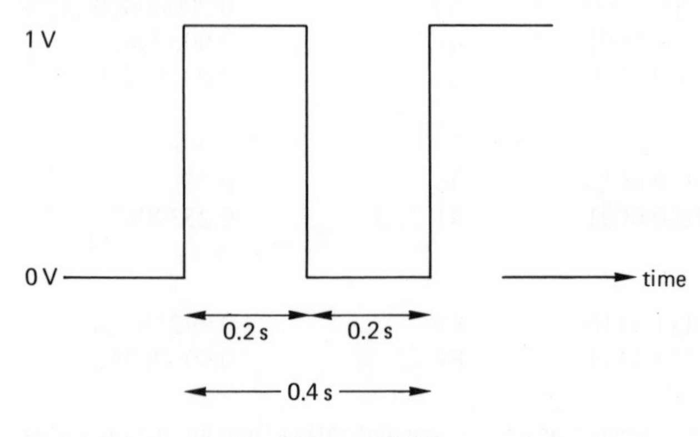

Figure 11.10

In the following, it is assumed that the address of the DAC and I/O port is 02_{16}. The ON and OFF periods of the waveform (the 'mark' and 'space' periods) are determined by a software time delay; for simplicity, the 0.2-s time delay program TIME 1 in table 8.6 is used for this purpose, the time delay subroutine starting address being 0110.

The square wave program is given in table 11.16; initially, the hex word 00_{16} is moved to the DAC for 0.2 s, resulting in an output of 0 V from the DAC for this period of time. Following this, the hex word FF is moved to the output port to give an output voltage of 0.99609375 V for 0.2 s. This process is continuously repeated to provide the required square wave output.

The minimum and maximum values of the square wave can be modified under software control. For example, the output voltage changes between 0.5 V and 0.75 V if the contents of location 0001 in table 11.16 are changed to 80_{16} and the contents of location 0008 are changed to CO_{16}. Also, the mark-to-space ratio can be altered by using time delay sequences of differing duration during the two half cycles of the waveform.

Table 11.16 Square wave program

Address	Machine code	Label	Mnemonic	Comment
0000	3E	BEGIN:	MVI A,00_{16}	; load (A) with zero
01	00			
02	D3		OUT 02	; output to DAC
03	02			
04	CD		CALL TIME 1	; 0.2s time delay
05	10			
06	01			
07	3E		MVI A,FF_{16}	; load (A) with maximum value
08	FF			
09	D3		OUT 02	; output to DAC
0A	02			
0B	CD		CALL TIME 1	; 0.2s time delay
0C	10			
0D	01			
0E	C3		JMP BEGIN	; recommence cycle
0F	00			
10	00			

Sawtooth (ramp) waveform generator

The ramp waveform shown in figure 11.11, which rises in a series of approximately 0.0039 V steps from 0 to 0.996 V in 256 steps, is generated by the program in table 11.17. During its operation, the program calls for a time delay subroutine labelled TIME T, which results in the output voltage being maintained for a time

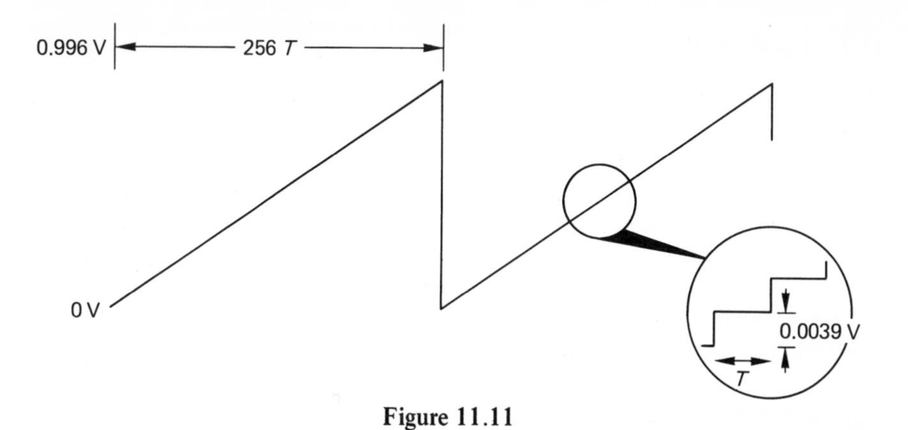

Figure 11.11

Table 11.17 Ramp wave program

Address	Machine code	Label	Mnemonic	Comment
0000	3E		MVI A,00_{16}	; load (A) with zero
01	00			
02	D3	WAVE:	OUT 02	; output zero to DAC
03	02			
04	CD		CALL TIME T	; time delay
05	VV			
06	UU			
07	3C		INR A	; increment (A)
08	C3		JMP WAVE	; next step
09	02			
0A	00			

duration T. Consequently, the time for the complete ramp is $256T$ s. The subroutine for TIME T is generally similar to that for TIME 1 (table 8.6), the starting address of the subroutine being given as $UUVV_{16}$, where UUVV is specified by the programmer.

The program operates as follows. A value of zero is initially output to the DAC for T seconds, after which this value is incremented by unity. Thus the analog output voltage commences at 0 V, and is incremented in steps of approximately 0.0039 V until it reaches 0.996 V (when the hex word FF_{16} is output to the DAC). The next INR A step in the program causes the value in the accumulator to become 00_{16}; the next OUTput instruction results in the output from the DAC being 0 V once more. The above program is repeated indefinitely. The periodic time of the waveform can be altered by modifying the TIME T subroutine.

11.12 Double Buffering

If a 10-, 12- or 14-bit DAC is used, care must be taken in designing the interface circuitry to prevent 'glitches' in the analog output signal. A **glitch** is a transient change in the output voltage for no apparent reason. This would occur, for example, in the case when data is loaded via the single buffered arrangement in figure 11.9 into a 10-bit DAC from an 8-bit microcomputer; the ten bits would have to be loaded in two stages, say the least significant 8 bits in one step and the most significant two bits in the second step, causing a glitch between the two steps.

One method of overcoming this defect is by **double buffering**. One circuit using this technique is shown in figure 11.12, and uses four output ports L_{L1}, L_{M1}, L_{L2} and L_{M2}. The program controlling these ports would first call for the least significant 8 bits to be latched into output port L_{L1}. Following this, the two most

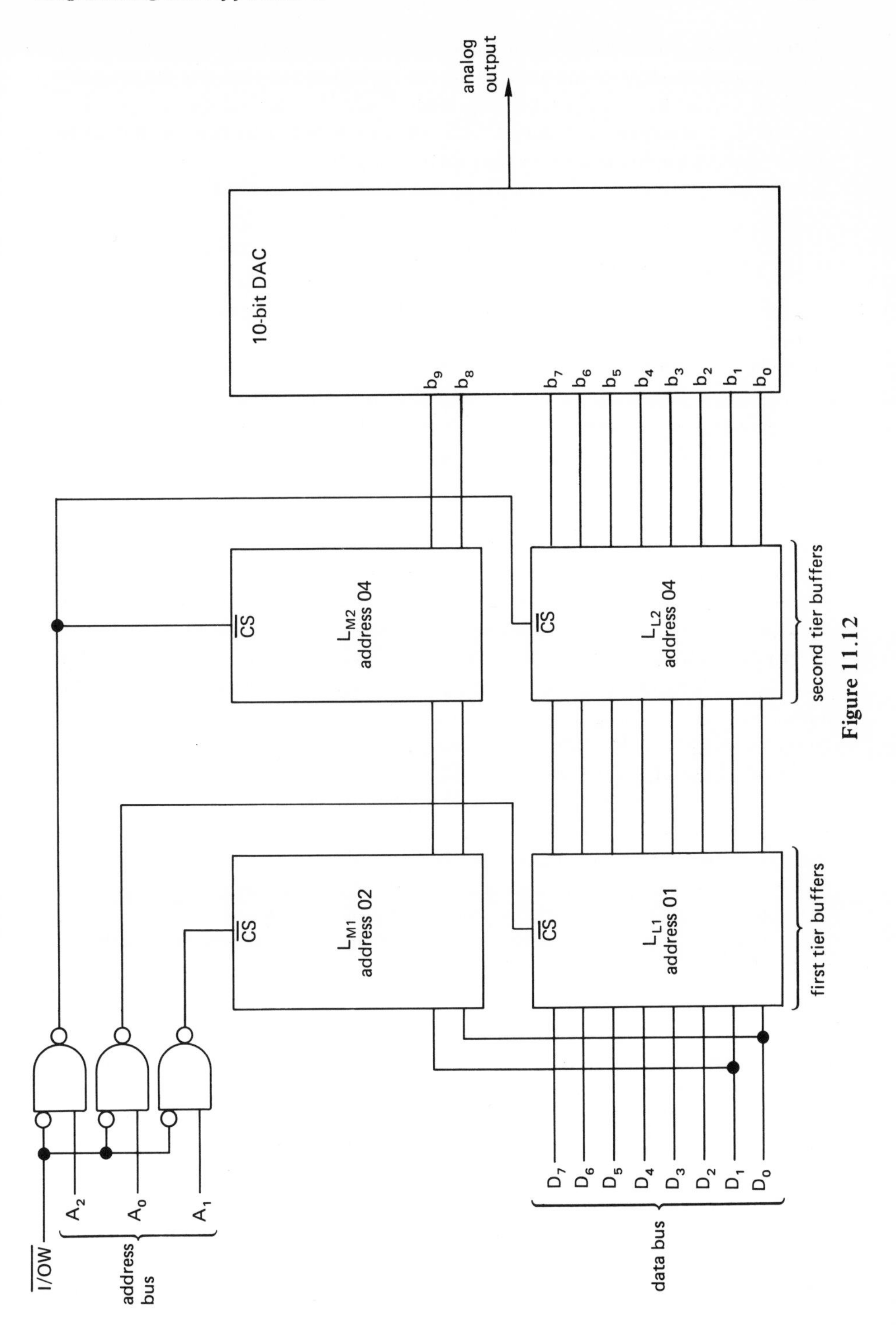

analog output
10-bit DAC
b9 b8 b7 b6 b5 b4 b3 b2 b1 b0
CS
L_M2
address 04
CS
L_L2
address 04
CS
L_M1
address 02
CS
L_L1
address 01
second tier buffers
first tier buffers
I/OW
A2
A0
A1
address bus
D7 D6 D5 D4 D3 D2 D1 D0
data bus

Figure 11.12

significant bits of the 10-bit word to be output to the DAC are latched into port L_{M1}. Thus all ten bits of the DAC word are latched in the first tier output ports. Next, all ten bits are simultaneously latched into the second tier output ports L_{L2} and L_{M2} when the instruction OUT 04 is executed. This ensures that all ten bits are applied simultaneously to the DAC.

11.13 A Simple Analog-to-digital Convertor (ADC)

The basis of a software driven **continuous balance** ADC is shown in figure 11.13a, together with a timing diagram in figure 11.13b. The ADC operates as follows. The CPU generates an analog ramp waveform using a program generally similar to that described in section 11.11; the essential difference between the ramp waveform in section 11.11 and the waveform required here is that in this case the ramp must 'run up' as fast as possible, that is, it is not necessary to introduce a time delay in the program between voltage steps. The rate at which the ramp rises depends on the number of instructions in the 'ramp' program; it may typically be 25 μs per voltage step, corresponding to a maximum ramp time of 255 x 25 μs = 6.375 ms.

In the circuit shown, a comparator is used to compare the ramp output voltage V_R with the unknown analog voltage V_U. When $V_R < V_U$, the output from the comparator is logic '0'; when $V_R \geqslant V_U$, the output from the comparator rises to logic '1'. The program controlling the ADC must therefore carry out the following steps.

(1) Set the contents of one of the CPU registers (say register B) to zero.
(2) Output the contents of register B to the DAC via output port 01 (see figure 11.13a).
(3) Input the comparator status via input port 01.
(4) Check if the comparator status is logic '1' and if it is, jump to step 6.
(5) Increment the contents of register B and jump to step 2.
(6) Conversion complete; transfer the contents of register B to the display device.

In the simplified program outlined above, register B was chosen to store the data which is output to the DAC chip. Initially, zero is output to the DAC, and the result of the comparison of V_R and V_U is input to the CPU via input port 01. The state of the signal on the data bus line D_0 is tested in the CPU by means of an ANI 01_{16} instruction, followed by a JNZ instruction; this causes program control to be transferred to step 6 if there is a match between V_R and V_U. If the comparator output is logic '0', the contents of register B are incremented and control is transferred to step 2, that is, V_R is increased in value and it is compared with V_U.

The loop is continuously cycled until a match is obtained between the DAC output and V_U, which is detected when the signal on D_0 = '1' in step 4. What the

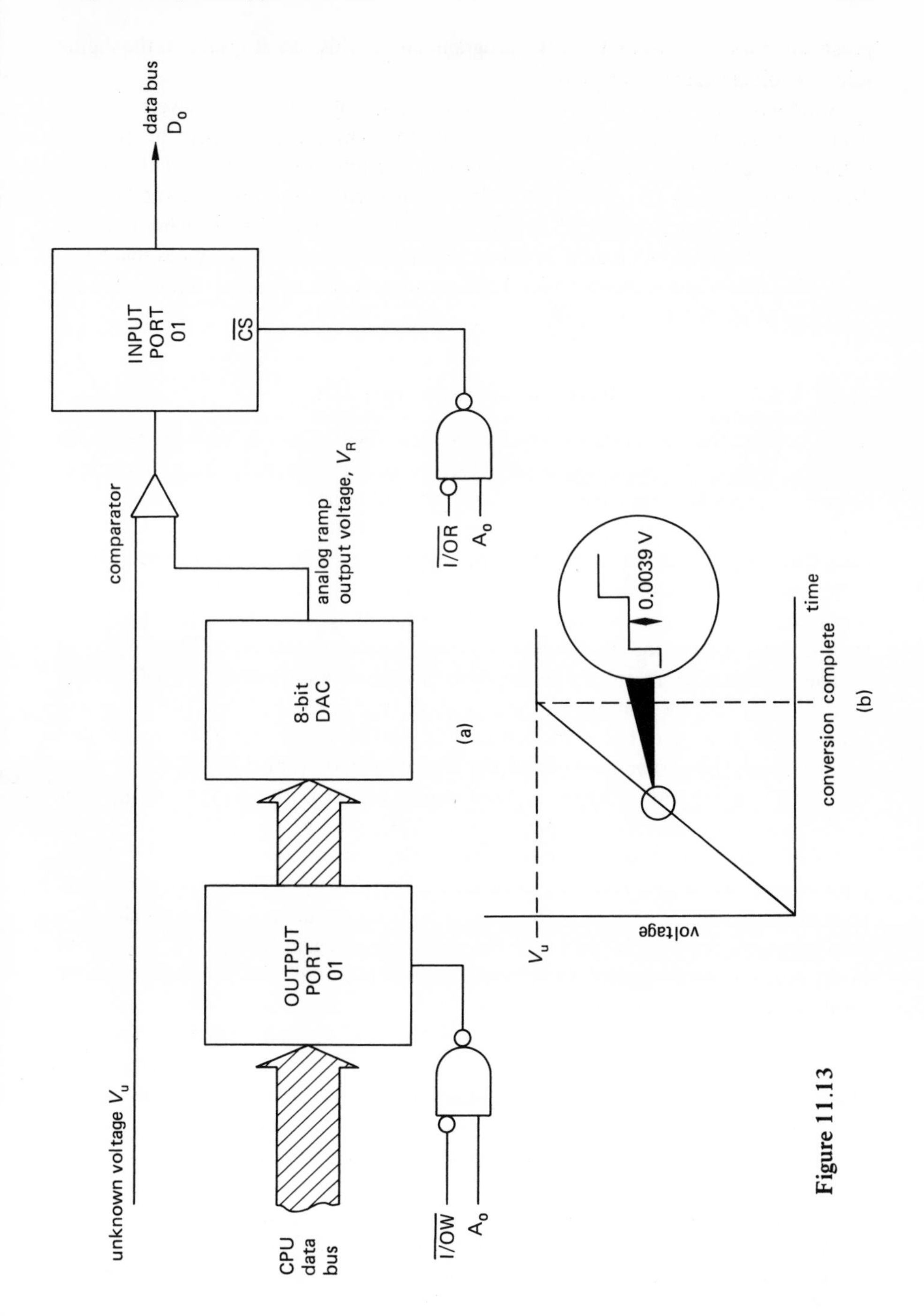
unknown voltage V_u
comparator
INPUT PORT 01
$\overline{CS}$
data bus D_o
8-bit DAC
analog ramp output voltage, V_R
OUTPUT PORT 01
CPU data bus
$\overline{I/OW}$
A_o
$\overline{I/OR}$
A_o
(a)
V_u
voltage
0.0039 V
time
conversion complete
(b)

Figure 11.13

processor does next depends on the program, and in this case it results in the digital value of voltage being displayed.

Although each step in the sample-and-test sequence takes only several tens of microseconds, it takes nearly 7 milliseconds to complete 255 passes of the loop (which is necessary in the case of a voltage at the limit of the range of the ADC). Thus the time taken to convert an analog voltage within the range of the DAC to its digital value lies between 25 μs and 7 ms. When compared with many systems, this is a relatively slow method of conversion. A faster method known as successive approximation is used where a very high conversion rate is needed. This method is described in section 11.14.

11.14 A Software-driven Successive Approximation ADC

In the case of the continuous balance ADC described above, the system may have to execute up to 256 sampling steps before balance is achieved. Using a technique known as **successive approximation**, balance is obtained after only a relative few sampling steps; if the analog signal is to be quantised into n bits, then the ADC need carry out only n sampling steps. That is, any analog voltage within the range of an 8-bit convertor is quantised after only 8 sampling steps. If a 10-bit convertor is used, the analog voltage is quantised after 10 sampling steps. However, the system *must* execute n sampling steps before the quantisation process is complete.

The hardware requirements of an 8-bit successive approximation ADC are the same as that for the continuous balance ADC (see figure 11.13a). The difference between the two types lies in the software as follows. In the case of the continuous balance type, the software results in the digital output from the CPU being incremented by an amount equal to the least significant bit during each sampling step. In the case of the successive approximation ADC, the software calls for a sequence of binary values to be used, commencing with the most significant bit (bit 7 in an 8-bit convertor); if this value is too small, a logic '1' is stored in bit 7 of the 'result' register in the CPU, and if it is too large a '0' is stored in bit 7 of the register. The next lower bit (bit 6) is added and if the result is once again too small a '1' is stored in bit 6 of the result register, and if too large a '0' is stored. This process is repeated until all eight bits have been tested.

The above process is illustrated for a 4-bit ADC (a 4-bit device is chosen for simplicity) which quantises a voltage in the range 0–15 V in 1-V steps (note: this results in a voltage of 4.4 V being quantised as 4V). Suppose that the DAC section of the circuit converts the least significant '1' into a 1-V step, then the sequence of steps in table 11.18 occurs when balancing an analog voltage of 13 V.

In the first step, the m.s.b. (that is, bit 3, which corresponds to a value of 1000_2 or 8 V) is converted into its analog equivalent and is compared with V_U. Since V_U (13 V) is greater than V_X (now 8 V), then a '1' is recorded in b_3 of the result store. Next, bit b_2 is set to logic '1' and its value is added to the result of the previous exercise. Once more the current value of V_X (12 V) is compared with V_U; since $V_U > V_X$, bit b_2 in the result register is set to logic '1'. Bit b_1 is next set to logic

Table 11.18 Analog-to-digital conversion of 13 V

Step	Test	Decision	Binary value stored
1	Is $V_U \geqslant 8$ V?	yes	$b_3 = 1$
2	Is $V_U \geqslant (8 + 4)$ V?	yes	$b_2 = 1$
3	Is $V_U \geqslant (8 + 4 + 2)$ V?	no	$b_1 = 0$
4	Is $V_U \geqslant (8 + 4 + 1)$ V?	yes	$b_0 = 1$

'1', which causes V_X to be 14 V; in this case $V_U < V_X$, and a '0' is recorded in b_1 of the result register. Finally, bit b_0 is set to logic '1' to give the new value of V_X as $(8 + 4 + 1) = 13$ V. The result of the comparison in this case is that $V_U = V_X$. Thus the result after four sampling tests is $V_U = 1101_2$ V $= 13_{10}$ V.

A flowchart for operating an 8-bit successive approximation ADC is given in figure 11.14 and a typical program is provided in table 11.19. The following registers are used in the program.

Register A (accumulator): I/O transfers and data manipulation
Register B: loop counter
Register C: result storage
Register D: test pattern used in sampling step.

Initially, registers A and C are cleared in order to remove irrelevant data, and 08_{16} is loaded into register B. Register B acts as a loop counter, and its contents are decremented at the end of each test sequence (each loop); the end of the test sequence is detected when the contents of register B are zero. Register C is used not only for the storage of intermediate results, but also the storage of the final result. The program listed in table 11.19 is terminated by a HALT instruction. (The terminating instruction in a user's program may differ from this, since it depends on what he wishes to do with the data on completion of the program.) The program is briefly described below.

Initially the hex pattern 80_{16} (that is, the m.s.b. of the word is logic '1') is output via output port 01 to the DAC (see figure 11.13). The DAC converts this into an analog voltage V_R, which is compared in a comparator with the unknown analog voltage V_U. An output of logic '0' from the comparator indicates that V_R is less than V_U, and a logic '1' indicates that V_R is either equal to V_U or is greater than it.

If V_R is greater than V_U, instructions 0011 to 0014 in table 11.19 result in the test bit being reset to zero, the modified result being stored in register C. If V_R is greater than V_U a jump is made to the point CONT in table 11.19 (the above instructions being ignored). Instructions 0015 to 0018 not only result in the test bit being rotated right but also cause the intermediate result to be transferred to the accumulator. Finally, the loop counter is decremented (instruction 0019) and

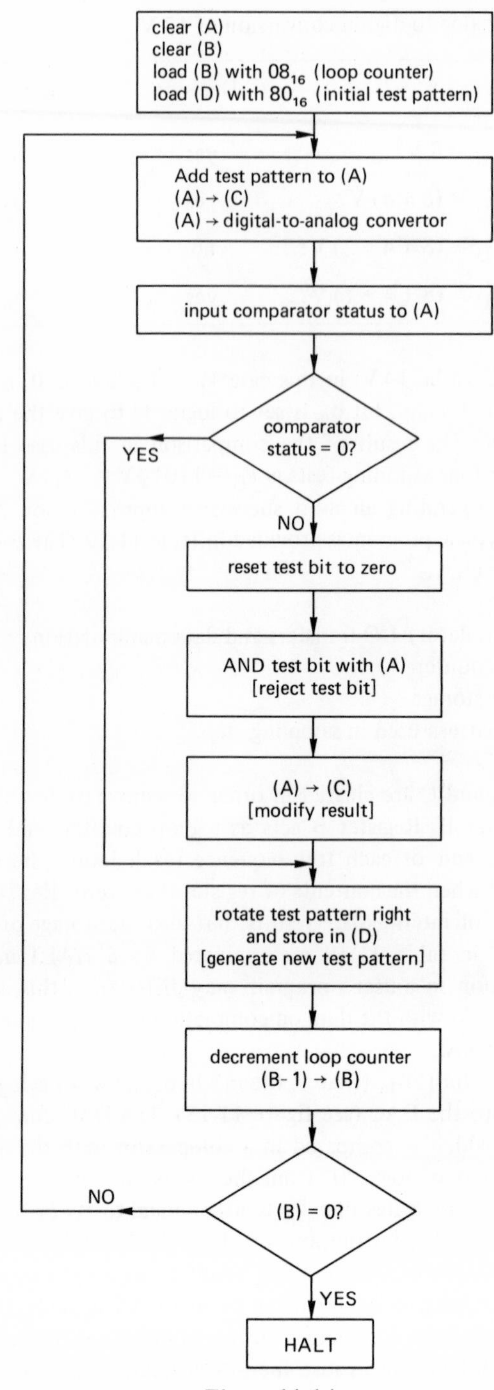

clear (A)
clear (B)
load (B) with 08_{16} (loop counter)
load (D) with 80_{16} (initial test pattern)
Add test pattern to (A)
(A) → (C)
(A) → digital-to-analog convertor
input comparator status to (A)
comparator status = 0?
YES
NO
reset test bit to zero
AND test bit with (A)
[reject test bit]
(A) → (C)
[modify result]
rotate test pattern right and store in (D)
[generate new test pattern]
decrement loop counter
(B- 1) → (B)
(B) = 0?
NO
YES
HALT

Figure 11.14

Table 11.19

Address	Machine code	Label	Mnemonic	Comment
0000	97		SUB A	; clear (A)
01	4F		MOV C,A	; clear (C)
02	06		MVI B,08_{16}	; load loop counter with 08_{16}
03	08			
04	16		MVI D,80_{16}	; load initial test pattern
05	80			; into (D)
06	B2	NEXT:	ORA D	; add old test pattern to new test pattern
07	4F		MOV C,A	; store result in register C
08	D3		OUT 01	; output (C) to DAC
09	01			
0A	DB		IN 01	; get comparator status
0B	01			
0C	E6		ANI 01_{16}	; mask out unwanted bits
0D	01			
0E	CA		JZ CONT	; jump if V_X less than V_U
0F	15			
10	00			
11	7A		MOV A,D	; transfer test pattern to (A)
12	2F		CMA	; set test bit to zero
13	A1		ANA C	; reject bit tested
14	4F		MOV C,A	; restore result in (C)
15	7A	CONT:	MOV A,D	; get test pattern
16	1F		RAR	; rotate test pattern right
17	57		MOV D,A	; put new test pattern in register D
18	79		MOV A,C	; transfer intermediate result to (A)
19	05		DCR B	; decrement loop counter
1A	C2		JNZ NEXT	; jump to NEXT if loop counter non-zero
1B	06			
1C	00			
1D	76		HLT	; halt

its contents are tested to check if the program loop has been executed eight times (as mentioned above, this is detected when the contents of register B are zero). If not, program control is returned to instruction 0006. The reader will note from a comparison of the flowchart and the program that the ORA D instruction (instruction 0006) effectively adds the test pattern to the contents of the accumulator; the

OR instruction can be used to replace an addition instruction, since a carry is not generated at any time by the addition process used here. When the program is complete, it is terminated by a HALT instruction.

11.15 Interfacing an 8-bit ADC Chip to a Microprocessor

Rather than use a software-driven ADC, the reader may prefer to use a commercially available ADC chip; this chip must then be interfaced to the CPU via I/O ports.

The basis of one form of 8-bit ADC chip is shown in figure 11.15. The analog voltage conversion process commences when the CPU sends out a pulse to the START terminal of the ADC; when this occurs, the DONE/$\overline{\text{BUSY}}$ line of the chip is driven low to indicate that the ADC is BUSY with the process of analog-to-digital conversion. Meanwhile, the successive approximation logic circuitry within the chip generates a sequence of test signals which are applied to an 8-bit DAC. The output voltage, V_X, from the DAC is compared with the unknown analogue input voltage V_U in the comparator section. When $V_X = V_U$, the conversion is complete,

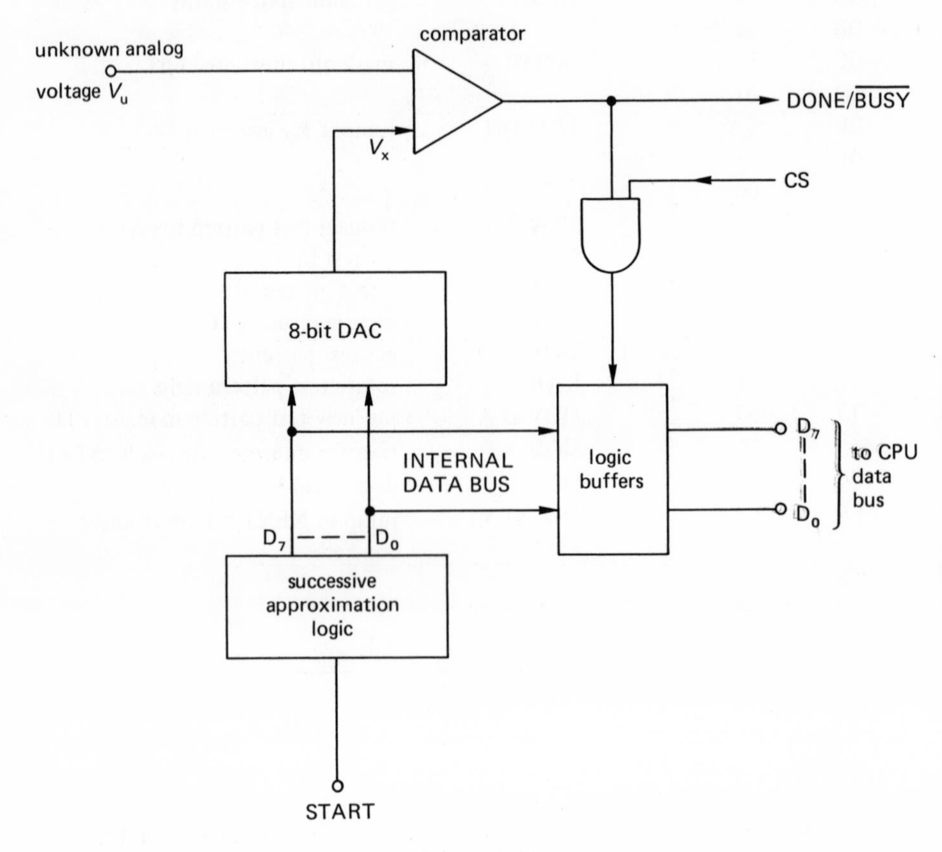

Figure 11.15

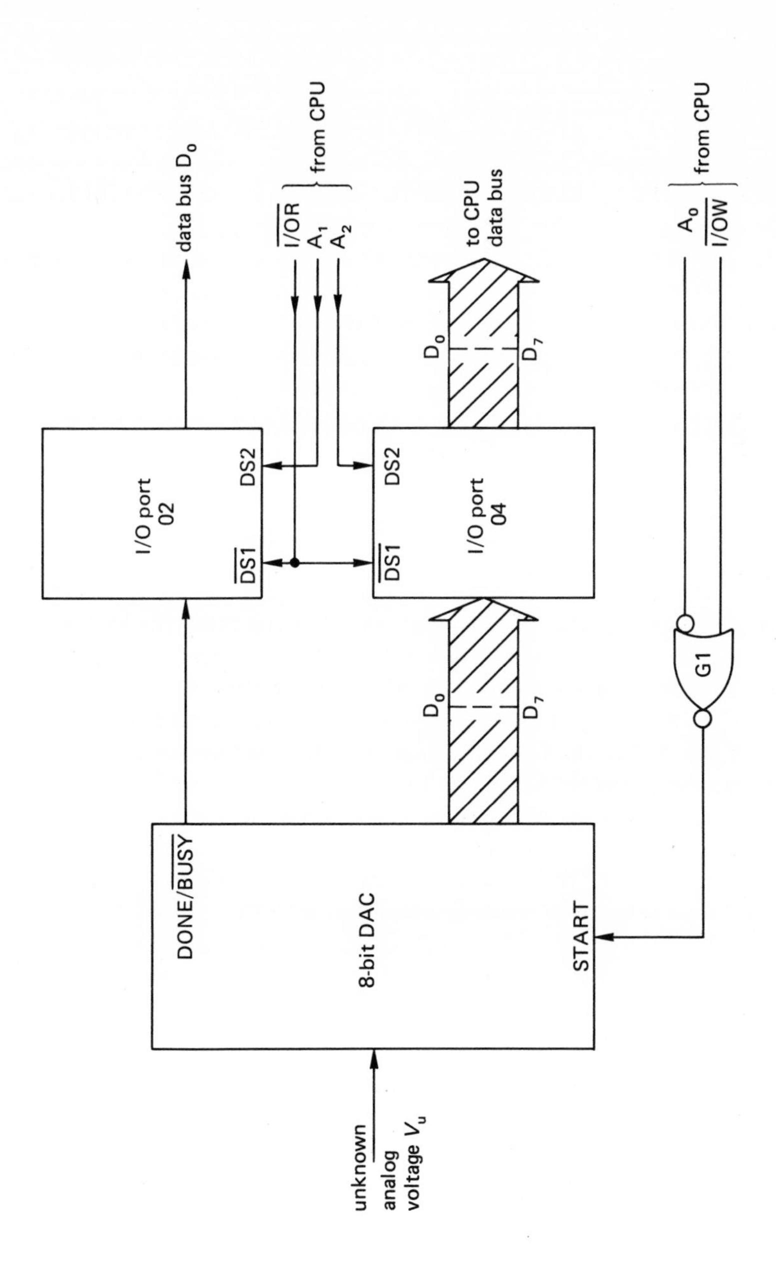
data bus D_0
$\overline{I/OR}$
A_1
A_2
from CPU
to CPU data bus
A_0
$\overline{I/OW}$
from CPU
I/O port 02
DS2
$\overline{DS1}$
I/O port 04
DS2
$\overline{DS1}$
D_0
D_7
D_0
D_7
G1
DONE/$\overline{BUSY}$
8-bit DAC
START
unknown analog voltage V_u

Figure 11.16

Table 11.20 Subroutine for the control of an 8-bit ADC

Address	Machine code	Label	Mnemonic	Comment
0200	D3	ADC:	OUT 01	; START conversion
01	01			
02	DB	BUSY:	IN 02	; input DONE/$\overline{\text{BUSY}}$ signal
03	02			
04	E6		ANI 01	; mask out unwanted bits
05	01			
06	CA		JZ BUSY	; jump if conversion incomplete
07	02			
08	02			
09	DB		IN 04	; conversion complete,
0A	04			; input data to CPU
0B	C9		RET	

and the DONE/$\overline{\text{BUSY}}$ line is driven high; this signal is used to enable the 3-state output buffers in the ADC chip. Thus a logic '1' on the DONE/$\overline{\text{BUSY}}$ line indicates to the CPU that the conversion is complete and that valid data is available.

One method of interfacing an 8-bit ADC chip to a microcomputer is shown in figure 11.16. The START pulse is derived from gate G1, and the DONE/$\overline{\text{BUSY}}$ signal is connected to the CPU via I/O port 02 to data bus line D_0. When the CPU senses that the DONE/$\overline{\text{BUSY}}$ line is high, the ADC is allowed access to the CPU data bus via I/O port 04. A typical program, written in the form of a subroutine, which controls the system in figure 11.16 is given in table 11.20. The I/O ports 02 and 04 in figure 11.16 are included in many commercial ADC chips, so that the practical circuit has a lower chip count than figure 11.16.

PROBLEMS

11.1 Write a sequence of instructions that evaluates the expression $A = B + C + D$. The values of B, C and D are stored in consecutive locations commencing at 0100_{16}, and the result is to be stored in location 0103_{16}.

11.2 Write a program that evaluates the expression $S = T + (U \times V)$. The values of T, U and V are stored in consecutive locations commencing at 0104_{16}, and the result is to be stored in location 0107_{16}.

11.3 Write a program that generates the 1's complement of the contents of memory location 0030_{16} and puts the result in location 0031_{16}.

11.4 Write a program that compares the contents of memory locations 0050_{16} and 0051_{16} and places the larger value in location 0052_{16}.

11.5 Write a program which generates the square of the number in memory location 0050_{16} and places the result in location 0051_{16}. (Note: the value stored in location 0050_{16} must be less than 16_{10}.)

11.6 A number is stored in memory location 0040_{16}. Write a program which adds the number to the bottom of a list if it is not already present in the list, otherwise the list remains unchanged. The length of the list is stored in location 0045_{16}, and the list commences at location 0046_{16}.

11.7 Write a program which sorts a list of unsigned binary numbers in decreasing order. The length of the list is stored in memory location 004F, and the list begins at location 0050_{16}.

11.8 Write a program for a traffic light controller, the lights at the junction being connected to output port 04 as follows

bit 0 North-South Red
bit 1 North-South Amber
bit 2 North-South Green
bit 3 East-West Red
bit 4 East-West Amber
bit 5 East-West Green

The sensors in the road are connected to the CPU data bus via input port 02 as follows

bit 0 North-South sensor
bit 1 East-West Sensor

The lamp sequence is Red, Red and Amber, Green, Amber, Red. The time between the Red and Red-and-Amber sequence is 20 s, and that between the Red-and-Amber and Green sequence is 5 s.

11.9 The traffic light control sequence in problem 11.8 is modified to use a real-time clock which generates an interrupt signal every second. Write a program which causes the CPU to generate the sequence of lights given in problem 11.8.

Further Reading

Microprocessors

Microprocessors and Microcomputers, E. Huggins (Macmillan Press)
Introduction to Microprocessors, L.A. Leventhal (Prentice-Hall)
8080A/8085 Assembly Language Programming, L.A. Leventhal (Osborne and Associates)
Microcomputer Design, C.A. Ogdin (Prentice-Hall)
Microcomputer-Based Design, J.B. Peatman (McGraw-Hill)

Digital Circuit Design

Digital Electronic Circuits and Systems, N.M. Morris (Macmillan Press)
Logic Circuits, N.M. Morris (McGraw-Hill)

Semiconductor Devices

Semiconductor Devices, N.M. Morris (Macmillan Press)

Solutions

2.1 (a) $6F_{16}$; (b) CE_{16}; (c) $4F_{16}$; (d) $5E_{16}$;
2.4 $2C5_{16}$
2.5 (a) 176_{16}; (b) 566_8
2.6 (a) 01111111; 10000000; (b) 11111110; 11111111;
(c) 01010101; 01010110; (d) 10101010; 10101011.
2.7 (a) 00010111; (b) 11101001
2.8 (a) 17_{16}; (b) $E9_{16}$
2.9 110010_2
2.10 10_2

3.1 AND (G1), OR (G2), NOR (G3), NAND (G4)
3.6 A = 3000_{16}; B = 3001_{16}; C = 3002_{16}; D = 3003_{16}; E = 3004_{16};
F = 3005_{16}; G = 3006_{16}; H = 3007_{16}

4.1 (a) output port; (b) $\overline{\text{I/OW}}$; (c) $E3FF_{16}$

5.5 (a) (i) 40_{16} or $C0_{16}$, (ii) $3F_{16}$ or BF_{16};
(b) (i) $6D_{16}$ or ED_{16}, (ii) 12_{16} or 92_{16}

6.4 ROM: X0XX, X2XX, X4XX or X6XX;
RAM: X9XX, XBXX, XDXX or XFXX

7.7 2400 baud; 218.2

8.2 2000_{16} (Note: since the stack register is decremented prior to every stack write, the first location which is used in the stack is $(2000 - 1)_{16} = 1FFF_{16}$.)
8.3 7000_{16}

Index

Absolute address 117
Accumulator 75
Accumulator I/O 129–34
Addition, BCD 32–4
  binary 24
  hexadecimal 26
  multi-precision 31
Addition program, 8-bit 215
  16-bit 216
Address 4, 11
Address bus 4, 70
Address field 167
Addressing mode (method) 191
  auto-indexed 198
  direct 192
  immediate 193
  indirect 192
  indexed 193
  register direct 197
  register indirect 197
  relative 194
  stack 198
Alphanumeric code 23
Analog-to-digital convertor 240
  continuous balance 240
  interfacing for 246
  successive approximation 242
    program for 245
AND gate 42
Architecture, of computer 75
Arithmetic and logic unit (ALU) 2
Arithmetic shift 204
ASCII code 23–5
Assembler program 187
Assembly 82
Assembly language 81, 187
Attached I/O 130, 136
Auxiliary carry flag 76

Baud rate 148
Bidirectional data bus 9
Binary-coded-decimal (BCD) code 20–2
Binary digit (bit) 6
Bit testing program 217
Branch instructions *see* Jump instructions
Buffer 49
Bus 4
  driver 59
Byte 6

Call instructions 157, 171
Carry flag 76
Central processing unit (CPU) 1, 10
Chip enable pin 7
Clock signal 51
Comment, in program 167
Comparison instruction 202
Compliment, binary 26, 28–30
Conditional jump instructions 212
Control bus 4, 70, 85
Control unit of CPU 1, 2

Data bus 4, 79
Data manipulation instructions 199
  arithmetic 199
  comparison 202
  logical 203
  shift and rotate 204
Data rate 150
Data transfer instructions 205
  I/O 210
  memory 205
  register 209
  stack 210
Debouncing, software 92
Decimal adjustment 33
Decoder 60–2
Dedicated microcomputer 90
Delay (time) subroutine 164, 231
Delimiter 168

Device enable pin 7
Device select pin 7
Digital computer 1
Digital-to-analog convertor 234
  waveform generation using 236
Direct memory access (DMA) 127
Division, binary 36
  non-restoring 37, 38
Double buffering 238
D-type flip-flop 55
Dynamic RAM 116

EBCDIC code 24
Effective address 193
Enable pin 7
EPROM 103, 107, 108
Error detection 22
EXCLUSIVE-OR gate 44, 50
Expanding I/O facility 151

FAMOS 108
Fan-out 49
Field structure (of assembler instructions) 167
Fixed-point notation 15
Flag 23, 52, 75
Flag register 23, 76
Flashing light program 223
Flip-flop 51
  D 55
  J-K 53
  S-R 52
Floating-point notation 16
Flowchart 81, 214
Framing error 148

General-purpose interface bus (GPIB) 153
Glitch 238

Half-carry flag *see* Auxiliary carry flag
Handshake 136-8
Hardware 4
Hewlett-Packard interface bus (HPIB) 153
Hexadecimal code 15, 17

IC (integrated circuit) 56
IEEE-488 interface bus 151-3
Input/output 129
  accumulator 130
  attached 136
  isolated 129
  memory-mapped 134
  standard 129
Input/output instructions 11, 82-4
Input/output port 7, 62-6, 129
Input port 5, 7, 79
Instruction register 13, 75
Instruction set 187
Integrated-injection logic (I^2L) 57
Interface standards 151-4
Interface unit 5
Interrupt 173
  multilevel 177
  priority 177
  vectored 178
Interrupt flag 175
Interval timer 186
Isolated I/O 129-34

Jump instructions 84

Keyboard scanning 91

Label, in program 167
Latch 52
LED (light-emitting diode) 97-102
LIFO store 158
Loadable monitor 91
Logic instructions 203

Machine code instructions 187
Macro 191
Maskable interrupt 175
Memory expansion 117
Memory map 120-7
Memory-mapped I/O 130, 134-6
Memory read, waveforms for 72
Memory write, waveforms for 73
Microinstruction 188
Microprogram 188
Monitor 91
MOS (metal-oxide-semiconductor) transistor 57
Multiplexing 58
Multiplication, binary 34
  program for 220
Multiword arithmetic 31

NAND gate 45, 48, 49
Negative logic 42
Negative number, binary 27
Nibble 6
Non-maskable interrupt 175

Non-volatile memory 104
NOR gate 46, 48, 50
NOT gate 45
Number conversion 18

Object code 82
Object program 82, 167, 187
Octal number 15, 20
On-chip I/O 130, 136
One's complement notation 28
Opcode 82, 187
Operand field, in assembly program 167
Operation code (op-code) 82, 187
OR gate 43
Output port 5, 7, 79
Overflow bit 26
Overrun error 150

Page 117
Parallel interface 5, 6
Parallel I/O 6
Parity 22
Parity flag 76
Polling 173
POP instruction 158, 162-4
Port 7
Positive logic notation 41
Priority encoder 66
Processor status word (PSW) 76
Program 2
Program counter 76
Programmable I/O (PIO) 129, 138-47
Program manipulation instructions 210
 conditional 212,
 subroutine 212
 unconditional 210
PROM 103, 106, 108
Pseudo-operation 191
PUSH instruction 158, 162-4

RAM (random-access memory) 4, 77, 89, 103, 112, 127
Real-time clock 186
Register 75
Relocatable program 195
Reset, microcomputer 88, 90
RES instruction 169
RET instruction 157
Rollover (keyboard) protection 97
ROM (read-only memory) 4, 77, 87, 103, 106, 108, 127
Rotate instructions 204-6
RS-232 interface, 153

S-100 interface 153
Serial data transmission 147-51
Serial I/O 6
Seven-segment display 97, 99-102
Shift instructions 204
Sign bit 27
Sign flag 76
Source program 82, 167, 187
Source statement 82
Stack 157
Stack pointer 158
Standard I/O 129
Static RAM 112
Status management instructions 212
Status register 23
Subroutine 156
 nested 168
Subtraction binary 29
 hexadecimal 30
Subtraction program 217
Successive-approximation ADC 242-6
Symbolic language 167

Three-state logic 57
Timer 185
Timing loop, software for 165
Traffic light program 227
Tri-state logic *see* Three-state logic
Truth table 40
TTL 57
Two-key rollover protection 97
Two's complement notation 28-30

UART 147

Vectored interrupt 176, 178
Volatile memory 104

Word length 6

Zero flag 76